D0040851

Waiting to Be Heard

Amanda Knox

Waiting to Be Heard

A Memoir

HARPER

www.harpercollins.com

FIRST EDITION

All photographs courtesy of the author unless noted otherwise.

Map copyright © 2013 by Springer Cartographics LLC.

Designed by Leah Carlson-Stanisic

Library of Congress Cataloging-in-Publication Data has been applied for.

ISBN: 978-0-06-221720-2

13 14 15 16 17   OV/RRD   10 9 8 7 6 5 4 3 2 1

For my family.

Contents

Prologue: December 4, 2009, Perugia, Italy *1*

Part One: PERUGIA

chapter 1   April–August 2007, Seattle, USA 5

chapter 2   August 30–September 1, 2007, Italy 15

chapter 3   September 2007, Perugia, Italy 25

chapter 4   October 2007 39

chapter 5   October 25–November 1, 2007 51

chapter 6   Morning, November 2, 2007,
Day One 65

chapter 7   Afternoon, November 2, 2007,
Day One 77

chapter 8   November 3, 2007, Day Two. 85

chapter 9   November 4, 2007, Day Three 97

chapter 10   November 5, 2007, Day Four 103

chapter 11   Morning, November 6, 2007, Day Five . . . 125

chapter 12   Evening, November 6, 2007, Day Five . . . 143

chapter 13   November 7, 2007. 153

chapter 14   November 8–9, 2007. 163

Part Two: CAPANNE I

chapter 15   November 10–13, 2007 173

chapter 16   November 9–14, 2007 181

Contents

chapter 17   November 15–16, 2007 197

chapter 18   November 2007 205

chapter 19   November 18–29, 2007 219

chapter 20   December 2007 229

chapter 21   January–May 2008 241

chapter 22   June–September 2008 259

chapter 23   September 18–October 28, 2008 271

chapter 24   October–December 2008 285

chapter 25   January–March 2009 289

chapter 26   March–July 2009 313

chapter 27   September 1–October 9, 2009 329

chapter 28   October 10–December 4, 2009 347

chapter 29   December 4, 2009 367

Part Three: CAPANNE II

chapter 30   December 2009–October 2010 375

chapter 31   November–December 2010 393

chapter 32   December 11, 2010–June 29, 2011 405

chapter 33   June 29, 2011 423

chapter 34   June 30–October 2, 2011 427

chapter 35   October 3, 2011 437

Epilogue: October 3–4, 2011 . . . . *447*

Author's Note *459*

Waiting to Be Heard

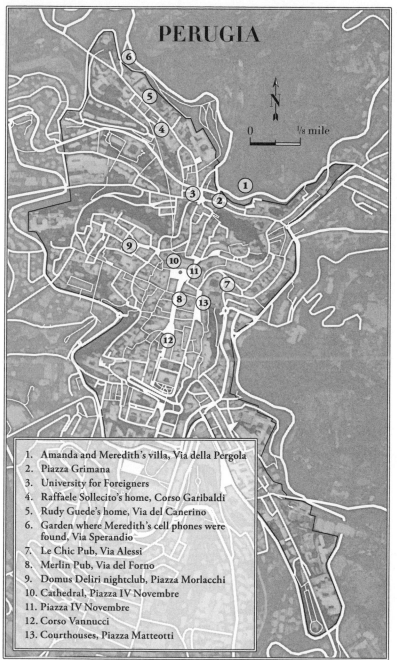

PERUGIA

N

0   ⅛ mile

1. Amanda and Meredith's villa, Via della Pergola
2. Piazza Grimana
3. University for Foreigners
4. Raffaele Sollecito's home, Corso Garibaldi
5. Rudy Guede's home, Via del Canerino
6. Garden where Meredith's cell phones were found, Via Sperandio
7. Le Chic Pub, Via Alessi
8. Merlin Pub, Via del Forno
9. Domus Deliri nightclub, Piazza Morlacchi
10. Cathedral, Piazza IV Novembre
11. Piazza IV Novembre
12. Corso Vannucci
13. Courthouses, Piazza Matteotti

Map copyright © 2013 Springer Cartographics LLC

December 4, 2009

Perugia, Italy

I walked into the ancient Perugian courtroom, where centuries of verdicts had been handed down, praying that a tradition of justice would give me protection now. I glanced at the large crucifix on the wall, crowning the judge's seat. Blue-capped guards surrounded me, propelling me forward. The room was packed with police officers, lawyers, and journalists, but it was unnervingly quiet. I saw family—my mom, dad, stepmom, stepdad, my sister Deanna—standing over to one side, mouthing, "I love you, I love you." My other sisters were too young to be allowed in the courtroom, but they were waiting for me just on the other side of the double doors.

The injustice was finally—almost—over.

Four minutes after midnight a bell rang once, and the court secretary announced, "*La corte.*" The judges, wearing black robes, and the jury, draped with the green, white, and red of the Italian flag, came somberly through the chamber door. They looked stubbornly above and beyond our expectant faces as they walked to their places. I was standing between my two Italian lawyers, gripping the hand of the bigger man, the one who had told me again and again over all these months, "Courage, Amanda, we need you to have courage. We will do the rest."

I took in a deep breath as the judge lifted the paper and began reading the articles on which I was being tried, quietly, monotonously.

Someone behind me wailed, "No!" a second before I heard the judge pronounce, *"Colpevole"*—"Guilty." Trembling, I slumped into my lawyer, who put his thick arm around me and pushed my face into his chest. Blood was pounding in my ears. I kept moaning, "No, no, no." I thought, *This is impossible, this is impossible, this is a nightmare, this can't be true, it's not fair, it's not fair.* People were everywhere, shouting for or against me. Hands reached out to me, touched me—I didn't know whom they belonged to. Over all the noise and confusion, I could hear my sister and mother sobbing.

My legs couldn't support me. The guards held me up by my armpits and carried me, crumpled, out of the courtroom. In the chaos of my shattered world, I never heard the judge sentence me: "Twenty-six years."

Done. It was done.

Part One

PERUGIA

Chapter 1

April–August 2007

Seattle, USA

M om sat next to me in our favorite tall-back booth. Dad slid in across from us. "What's this about?" he asked. I couldn't believe the three of us were actually doing this.

Eating salads with my parents doesn't sound like a big deal, but it was for me. I'm sure that for them it was hugely uncomfortable. I was nineteen, and as far back as I could remember I'd never seen my parents sit at the same table, much less share a meal. I was a year old and my mom was pregnant with my sister Deanna when she and my dad split up. They had rarely talked to each other since, even on the phone. Proof of how much they both loved me was this reunion at the Eats Market Café in West Seattle. Mom picked at her fingernails. Dad was businesslike. All smiles were for me.

The biggest testament to my parents' love for Deanna and me was how they'd handled their divorce. They bought houses two blocks apart to give us the benefits of a two-parent family and the gift of never feeling pulled between them. I never once heard either criticize the other. But they were invisible to each other, whether separated by two blocks or two rows at a school play. At soccer games, both cheered on the sidelines buffered by a line of other parents.

The permanent divide meant that when I had news to tell I always had to do it twice. Bringing my parents together this one time was my way of saying: this is the most important decision of my life so far. It was a drumroll to let them know that I was ready to be on my own.

As always, I had gone to my mom first. She's a free spirit who believes we should go where our passions lead us. When I told her mine were leading me 5,599 miles away from home, to Perugia, Italy, for my junior year of college, her unsurprising response was "Go for it!"

Mom was born in Germany and moved to Seattle as a child, and my grandmother, Oma, often spoke German to Deanna and me when we were growing up. It wasn't until my freshman year in college that I realized I had a knack for languages and started playing around with the idea of becoming a translator. Or, if only, a writer. When it came time to decide where to spend my junior year, I thought hard about Germany. But ultimately I decided to find a language and a country of my own—one my family hadn't already claimed. I was sure that would help me become my grown-up self—whoever that was.

Germany would have been the safer choice, but safety didn't worry me. I was preoccupied by independence. I trusted my sense of responsibility, even if I sometimes made emotional choices instead of logical ones—and sometimes they were wrong.

If I really wanted to become a translator, Spanish or French would have been a more practical choice than Italian. But everyone took Spanish, and I didn't feel connected to French. My fascination with Italian culture went back to middle school, when I studied Latin and learned about Roman and Italian history. I loved Italy even more when I was fourteen and saw it close up, on a two-week trip with my mom and her family. My Oma, aunts, uncles, stepdad, Deanna, and I piled into two minivans and drove through Germany and Austria to visit relatives and celebrate Oktoberfest in Munich before heading south into Italy, to see Pisa, Rome, Naples, Pompeii, and the Amalfi Coast. The history I'd studied became real when we visited the Coliseum and the ruins of Pompeii. I remember pointing out things to my family and babbling about factoids I'd stored away, so much so that they nicknamed me "the tour guide." I was charmed by the narrow, cobblestone streets and the buildings rooted into the earth that were so different from what I was used to in Seattle. It was a month and a half after 9/11, and all the Italians we met were warm and sympathetic. I came away thinking Italy was a welcoming, culturally and historically rich country.

As a sophomore in college, I signed up for Italian 101. Then, when I found out that the University of Washington hosted a summer creative writing program in Rome, taught in Italian, it felt like kismet. It combined everything I was looking for. Step one was to master Italian and immerse myself in the culture for nine months in tiny Perugia. Then I'd be ready to take on Rome in June.

Now I had to convince my dad. He's a linear thinker who works in finance. He's into numbers and planning. As practical and organized as he is, he'd have a lot of questions. So I approached him armed preemptively with the answers.

I also had another mission when I'd set up the two-parent

lunch. I wanted to show my dad that I loved him and my mom equally. While I was asking him to trust me in Perugia, I was also asking to be forgiven.

During my first two years in college, I'd gotten better at seeing things from other people's perspectives. I started mentally cataloguing the times I'd been selfish. A big one was how I'd treated Dad when I was a teenager.

Growing up, I officially spent every other weekend with him. Dad wasn't a micro-parent; he left all the day-to-day details to my mom. When I had a decision to make I turned to her. She would lay out the options and encourage me to make the choice myself. Dad wasn't part of the process.

The house he shared with his second wife, Cassandra, was theirs, not mine. When my half sisters, Ashley and Delaney, were born, Deanna and I were relocated from our shared bedroom in that house to pullout couches in the playroom. My real home was the one with my room, my sister, my mom, and her second husband, Chris. That's where I felt most myself. Mom let us wear whatever we wanted and build forts in the backyard. On rainy days, it was Deanna's and my job to redirect the lost slugs that pilgrimaged into the dining room beneath the backdoor. Dad required us to use drink coasters, arrange the movie cassettes and CDs in alphabetical order, and wear matching outfits.

At fourteen, I told Dad I was too busy with my extracurricular activities and friends to stay over with him anymore. The truth was that I was uncomfortable with the awkward divide between my life and his, so I widened the gap between us. Now I wanted to close it.

As I began researching programs in Italy, I realized that having my dad's support was fundamentally important to me. I'd never rehearsed any part in a play as hard as I had this conversation in my head. I wanted my dad to be impressed. I wasn't at all sure what I would do if he said no. Once we were seated, I couldn't wait a second longer. I started making my case even before the waiter brought us menus.

"Dad," I said, trying to sound businesslike, "I'd like to spend next year learning Italian in a city called Perugia. It's about halfway between Florence and Rome, but better than either because I won't be part of a herd of American students. It's a quiet town, and I'll be with serious scholars. I'll be submerged in the culture. And all my credits will transfer to UW."

To my relief, his face read receptive.

Encouraged, I exhaled and said, "The University for Foreigners is a small school that focuses only on language. The program is intense, and I'll have to work hard. The hours I'm not in class I'm sure I'll be in the library. Just having to speak Italian every day will make a huge difference."

He nodded. Mom was beaming at my success so far.

I kept going. "I've been living away from home for almost two years, I've been working, and I've gotten good grades. I promise I can take care of myself."

"I worry that you're too trusting for your own good, Amanda," he said. "What if something happens? I can't just make a phone call or come over. You'll be on your own. It's a long way from home."

Dad has a playful side to him, but when he's in parent mode he can sound as proper as a 1950s sitcom dad.

"That's the whole point, Dad. I'll be twenty soon, and I'm an adult. I know how to handle myself."

"But it's still our job to take care of you," he said. "What if you get sick?"

"There's a hospital there, and Aunt Dolly's in Hamburg. It's pretty close."

"How much is tuition? Have you thought about the extra costs involved?"

"I've done all the math. I can pay for my own food and the extra expenses," I said. "Remember I worked three jobs this past winter? I put almost all of it in the bank. I've got seventy-eight hundred dollars saved up."

Dad wove his fingers together and set them, like an empty basket, on the table. "How would you get around?"

"The university is right downtown, and there's a city bus," I said. "And Perugia is small. It's only about a hundred and sixty thousand people. I'm sure I'll learn my way around really fast."

"How will you stay in touch with us?"

"I'll buy an Italian cell phone, and I'll be on e-mail the whole time. We can even Skype."

"Will you live in a dorm?"

"No, I'll have to find my own housing, but I'm sure I can get a good apartment close to campus. I checked with the UW foreign exchange office—they say the University for Foreigners will give me a housing list when I get there. I'd really like to live with Italians so I can practice speaking the language."

I didn't know what my father would think, but I was pretty sure we'd be going back and forth for weeks no matter what. To my astonishment, he said yes before I'd picked up my fork.

"I'm proud of you, Amanda," he said. "You've worked hard and saved a lot of money. I can tell how much this means to you."

I knew I'd be just one of about 250,000 American college kids heading abroad in the fall, but this was the most momentous step I'd taken in my life so far. And I was unusual among the people I knew—most of my UW friends weren't studying abroad. I felt exceptional. I felt courageous. I was meeting maturity head-on. I'd come back from Italy having evolved into an adult just by having been there. And I'd be fluent in Italian.

———————

This year overseas would be the first time I'd ever really been on my own. During senior year at my Jesuit high school, Seattle Prep, almost all my friends sent applications to schools hundreds of miles from home. Some even wanted to switch coasts. But I knew that I wasn't mature enough yet to go far away, even though I didn't want to miss out on an adventure. I made a deal with myself. I'd go to the University of Washington in Seattle, a bike ride from my parents' houses, and give myself a chance to season up. By the time high school graduation came around, I'd already started looking into junior-year-abroad programs.

Most of my high school class had been more sophisticated than I was. They lived in Bellevue, a decidedly upscale suburb with mansions on the water. Their neighbors were executives from Boeing, Starbucks, and Microsoft.

I received financial aid to attend Prep and lived in modest West Seattle, not far from my lifelong friend Brett. I was the quirky kid who hung out with the sulky manga-readers, the ostracized gay kids, and the theater geeks. I took Japanese and sang, loudly, in the halls while walking from one class to another.

Since I didn't really fit in, I acted like myself, which pretty much made sure I never did.

In truth I wouldn't have upgraded my lifestyle even if I could have. I've always been a saver, not a spender. I'm drawn to thrift stores instead of designer boutiques. I'd rather get around on my bike than in a BMW. But to my lasting embarrassment, in my junior year, I traded my friends for a less eccentric crowd.

I'd always been able to get along well with almost anyone. High school was the first time that people made fun of me or, worse, ignored me.

I made friends with a more mainstream group of girls and guys, attracted to them by their cohesiveness. They traveled in packs in the halls, ate lunch together, hung out after school, and seemed to have known each other forever. But in pulling away from my original friends, who liked me despite my being different, or maybe because I was, I hurt them. And while my new friends were fun-loving, I was motivated to be with them by insecurity. I'm ashamed for not having had the guts to be myself no matter what anyone thought.

This didn't change who I was. Like most teenagers, I was too well aware of my flaws. I felt lumpy in my own skin. I was clumsy with words, and I knew I was way too blunt. I'd do things that would embarrass most teenagers and adults—walking down the street like an Egyptian or an elephant—but that kids found fall-over hilarious. I made myself the butt of jokes to lighten the mood. The people who loved me considered my kookiness endearing. My family and friends would shake their heads good-naturedly and sigh, "That's Amanda."

Soccer is where the boundaries fell away. I was good at it, and that always allowed me to feel on par with others.

In college I finally found my footing off the field. I stayed in touch with Brett and met a small group of intelligent and off-beat students at the university's climbing wall and in my dorm. I dated a Mohawked, kilt-wearing, outdoorsy student named DJ. My next-door neighbor was a girl from Colorado named Madison. She and I grew close and looked up to each other. She wasn't like most of the students. She didn't play sports, drink, smoke, or go to parties. She was a conflicted Mormon and a musician, majoring in women's studies and photography. I kept her company at night in the campus dark room. She encouraged me to be myself.

Most of my other friends were male. We played football, jammed on the guitar, talked about life. After we smoked pot we would choose a food category—burgers, pizza, gyros, whatever—and wander around the neighborhood until we found what we considered the best in its class.

As I got ready to leave for Perugia, I knew I hadn't become my own person yet, and I didn't quite know how to get myself there. I was well-meaning and thoughtful, but I put a ton of pressure on myself to do what I thought was right, and I felt that I always fell short. That's why the challenge of being on my own meant so much to me. I wanted to come back from Italy to my senior year at UW stronger and surer of myself—a better sister, daughter, friend.

While I was figuring out what I would need in Italy—my climbing gear, hiking boots, and a teapot were among the essentials—old friends from high school and new friends from college dropped by with well-wishes, little presents, and gag gifts.

I received a blank journal and a fanny pack and tins of tea. Funny, irreverent Brett brought me a small, pink, bunny-shaped vibrator. I was incredulous; I had never used one.

"Until you meet your Italian stallion," Brett said, handing it to me. She winked.

I laughed. The bunny was typical Brett. She liked to tease that I was regrettably behind everyone else. In high school she tried to coax me into straightening my hair and wearing makeup. I tried the first and thought it was okay. I tried the second and felt like an imposter. Her newest cause was to convince me to give casual sex a chance. I'd heard the same thing from other friends. It seemed to make some sense. I yearned to break down all the barriers that stood between me and adulthood. Sex was a big one—and the one that scared me the most. I'd bloomed late and didn't kiss a guy until I was seventeen. I lost my virginity after I started college. Before Italy, I'd had sex with four guys, each in a relationship I considered meaningful, even though they had turned out to be short-lived.

I left for Italy having decided I needed to change that. For me, sex was emotional, and I didn't want it to be anymore—I hated feeling dependent on anyone else. I wanted sex to be about empowerment and pleasure, not about *Does this person like me? Will he still like me tomorrow?* I was young enough to think that insecurity disappeared with maturity. And I thought Italy would provide me the chance to see that happen.

On the day I was leaving—in a rush to get to the airport and without a single thought—I tossed Brett's pink bunny vibrator into my clear plastic toiletry bag.

This turned out to be a very bad idea.

Chapter 2

August 30–September 1, 2007

Italy

I t seemed totally harmless. My sister Deanna and I were taking the train from Milan to Florence. Our seatmate, Cristiano, was tan, blond, and wearing a tank top that showed off his lean, muscular frame. He had the chiseled good looks of a California beach bum and the alluring accent of an Italian—a combination I found incredibly attractive. His English was even more limited than my rookie Italian. We made up for the gap with lots of gestures and smiles. Flirting, I realized, is a universal language.

Rolling past bright green fields, I wondered if Cristiano thought I was cute—the word I used to describe myself back then. The leap to beautiful or sexy was too huge for me to make. In my mind, if that ever happened, it would only be after I'd grown into a sexually confident woman. When that would be, I couldn't guess. But I was also increasingly aware, as I noticed Cristiano stealing glances in my direction, that some men saw in me something they found appealing.

Cristiano was going to Rimini, a seaside resort known for discos. Deanna and I were spending Thursday night in Florence and leaving for Perugia early the next morning. I was giddy. After reading about Perugia for months, I was finally going to see it. Deanna and I were giving ourselves two days to find me a place to live near the Università per Stranieri—the University for Foreigners—where my classes would start on October 1. Then my sister and I would take the train to Hamburg to vacation with our German cousins.

I thought of this side trip as the unofficial kickoff of a new phase in my life. I'd hang out with Italians and make new friends. Just being in a new environment would stretch me as a person.

We all got off the train in Florence. Deanna and I had planned the stopover, but Cristiano missed his bus to Rimini and got a room in our hotel. The three of us sat outside and shared a large tomato, mozzarella, and basil pizza and a carafe of wine. By then it was so obvious that Cristiano and I were into each other that I'm sure Deanna felt like an add-on. As soon as we finished eating, she announced that she wanted to go to bed and left us alone. Cristiano and I were walking around the city arm in arm when he said, "Hey, do you like *spinelli*?"—"joints."

"Yeah, do you have any?" I asked.

We shared a joint, and then, high and giggly, we went to his hotel room. I'd just turned twenty. This was my first bona fide one-night stand.

I'd told my friends back home that I couldn't see myself sleeping with some random guy who didn't matter to me. Cristiano was a game changer.

We didn't have a condom, so we didn't actually have intercourse. But we were making out, fooling around like crazy, when, an hour later, I realized, *I don't even know this guy.* I jumped up, kissed him once more, and said good-bye. I went upstairs to the tiny room Deanna and I were sharing. She was wide awake, standing by the window. "Where have you been?" she asked. "I didn't know where you were or if you were okay."

She was right.

I tried to make it up to her by waking at dawn and racing around Florence taking snapshots of us funny-posing and face-making on the Ponte Vecchio and in front of the copy of Michelangelo's *David* and the Fountain of Neptune. It kind of worked.

We boarded the first train of the day for Perugia and arrived in my new town while the early morning was still hazy. Our mom's cousin Dolly, who'd booked all our reservations, had instructed us to catch a bus to our hotel. A lifelong resident of Germany, Dolly, whom I call "aunt," had traveled all over Europe and would know what to do. But I'm a walker, and it seemed ridiculous to take a bus. Besides, we didn't have a clue which one, and my Italian wasn't good enough to ask or to understand the answer. "Let's walk," I told Deanna. "It will help us get our bearings."

That was my second mistake in twelve hours.

Downtown Perugia is on top of a steep hill, and the train station is at the bottom. The climb up the stairs from the platform to the station left us breathless. I assumed the hotel wasn't far, and when we passed a kiosk, I bought a map. I found our hotel at the edge of the foldout. I figured the distance was doable, even though we were carrying backpacks filled with clothes and books in the late summer heat.

I led Deanna across a trestle over the train tracks and up a steep, winding road. As we walked, we looked out at cypress trees and olive groves, church steeples and terracotta houses. I breathed in the city's earthiness. I was already feeling like I owned the view. *Soon*, I thought, *this will be as familiar to me as the Space Needle in Seattle.*

After we slogged for an hour and a half, the sidewalk unexpectedly ended. The terrain grew even hillier, and when we came to a ramp for the highway, we had no choice but to walk alongside it. Tall, dry grass scratched our legs, and in no time we were dotted with bug bites. The trek was turning out to be at least twice as far and four times more rugged than I'd expected. The sun was high in the sky.

Sweating, miserable, and close to tears, Deanna finally said, "Amanda, this cannot be the right way." She lay down on top of her overstuffed backpack.

"You look like a turtle that's flipped over onto its shell," I said, trying to lighten our situation.

"I can't walk anymore," she said. "What are we going to do?"

I was still pretty sure I was right about our direction. Sitting down next to her, I pointed. "Here's where we are on the map. It'll get easier at the top of this rise. Then I'll find somebody who can tell us the way."

But before we got to the crest, a nondescript car pulled onto the shoulder. The driver looked perhaps a few years younger than my dad. I had no idea what he was asking or saying, but I'm sure he could tell we were lost Americans. Trying to communicate, we looked like we were playing roadside charades. But between his sparse English and my slight Italian vocabulary, we found two of the few words we had in common: *Holiday Inn.* He pointed to his car and traced his finger along the full length of our map, offering us a ride.

I'm trusting by nature—too trusting, as my dad had said—and I just assumed our driver was a decent guy. Really, what choice did we have? It's not like we could turn around. I was so relieved to find someone who knew how to get us to our hotel that I was happy to take a risk.

"*Grazie,*" I said.

I rode shotgun and did all the talking. On the off chance that he did anything crazy, I'd be the buffer between him and Deanna. As the oldest, I automatically reacted this way to any possibly dicey situation that included a sibling. I also felt safer when I had the illusion of being in control. Now, looking back, I see that I had a ridiculous amount of unwarranted self-confidence. Why did I assume I knew the way to a hotel in a country I'd been in once, years before, and a city I'd never been in at all? I hadn't been in a physical fight in my life. What could I have done to protect Deanna if the ride had gone wrong?

Fortunately, my take on our driver was better than my grasp of kilometers. After exiting the freeway, he made a series of sharp turns—there were no street signs—while cocking his head toward me to make small talk. I gathered that he either owned a disco or was inviting us to go dancing. I understood when he asked, "*Disco stasera?*"—"tonight." For ten tedious minutes I smiled and said no—the same word in English and Italian. Our driver wore an expression of cheerful, you-can't-blame-a-guy-for-trying resilience when he dropped us at our hotel.

By the time we'd checked in and left our bags in our room, we'd lost four crucial hours of apartment hunting. It could have been a lot worse. Done with walking, we took a bus back to town, where I bought a cheap mobile phone with prepaid minutes. Deanna and I stopped at a coffee bar on the main drag, spent five minutes trying to describe a mocha (*espresso* + *latte* + *cioccolato!*) to the good-humored barista, and I searched Perugia's classifieds for an apartment to rent—without success. My anxiety shot up. This trip had begun awkwardly. I wasn't superstitious, but I hoped it wasn't a sign of the way my year in Italy would go.

Deanna and I walked down a steep cobblestone street to the university and went into the ornate administration building. If I couldn't snag a place to live, at least I could find out how to register. Dozens of flags waved on poles on the balcony above the entrance, making the building look like a scaled-down United Nations. The students attending the university came from as many different countries as there were flags. As we were leaving, I saw a skinny brunette who looked a little older than I was. Wearing super-short cut-offs and a yellow tank top, she was taping a sign on a wooden railing crowded with all kinds of notices. She looked like a student, and I could see a phone number on her sign. I grabbed at the possibility. "Do you have an apartment to lease?" I asked tentatively in English. She answered—also in English, luckily—that she and her best friend were subletting two rooms in their rented house.

"How far is it?" I asked.

"It's right here, down this lane—two seconds," she answered. "Do you want to see it?" I couldn't believe that a likely solution to my biggest worry was standing right in front of me.

Her name was Laura Mezzetti, and I liked her immediately. Deanna and I followed Laura across the tree-lined piazza and past a series of redbrick high-rises. We were practically sprinting from one busy street to an even busier one. We crossed an intersection and came to a tall iron gate. Laura stopped and swung it open. We stood at the end of a driveway in front of a cream-colored stone villa with a terracotta roof fit for a fairy tale. It was on top of a hill that sloped down to a tangled, untended garden. I was flabbergasted. A villa in the middle of downtown!

"No way," I whispered to Deanna. "This is too perfect."

"The top floor is ours," Laura said. "The basement is rented by a group of guys—students."

Laura and her roommate, Filomena Romanelli, led us through the kitchen/living area. The house had four bedrooms, two bathrooms, and a terrace. One of the available bedrooms faced the driveway, with just a sliver of the valley view. The room next to it was slightly larger and had a picture window looking out on the countryside. Both cost the same, but I liked the smaller room better. It had everything I hoped for—a bed, a desk, a wardrobe, and a cozy feeling. The rent—three hundred euros, or just over four hundred dollars at the time— seemed expensive, but the place was close to the university, and a *villa*. It was worth it.

No. 7, Via della Pergola also felt like a happy place, probably because Laura and Filomena had a go-with-the-flow attitude toward life. "We go to work, we come home, we watch soap operas, we cook dinner, we hang out with friends," Laura said.

They were both in their late twenties and working at law firms. Laura was offbeat, with multiple piercings in both ears. Filomena seemed more girly but, like Laura, relaxed—a little bit of a hippie—and really funny. They reminded me of my friends in Seattle. I felt it was a good fit.

Laura's English was better than Filomena's. When she asked me about myself, I told her that I played the guitar but hadn't been able to bring mine to Perugia.

"Oh, I have one," Laura said. "You can use it anytime."

And when I said I did yoga, she replied, "Wow, can you teach me? I've always wanted to learn."

"You'll love the jazz and chocolate festivals," Filomena added. She offered Deanna and me fresh figs from the garden.

They said I wasn't the first roommate they'd interviewed. A guy they called "totally uptight" was interested in renting, until he found out they smoked—cigarettes and marijuana. "Are you okay with that?" Filomena asked.

"I'm from Seattle. I'm laid back," I answered. "I don't smoke cigarettes, but I'll share a joint." A few minutes later they rolled one and passed it around. I inhaled deeply and relaxed. I felt so lucky—and capable. Six thousand miles from home and without the help of my mom or dad, I'd organized the next chapter of my life. I'd found this amazing place to stay, and I would get to live as a local—Laura, Filomena, and their four downstairs neighbors were all Italian.

"I love it," I said. "I'll bring my deposit tomorrow. As soon as I can get to an ATM." Before we left, Deanna took a picture of Laura, Filomena, and me at the front door, smiling and with our arms around each other's backs.

Mission accomplished, Deanna and I left for Hamburg to stay with Aunt Dolly. I figured that when I returned in mid-September the last bedroom would be rented. Laura and Filomena had said they'd prefer another female but cared most about finding someone easygoing who would fit in. I liked them so much that I knew I'd get along with anyone they chose.

About a week after I got to Germany, Filomena and Laura e-mailed me that a British exchange student named Meredith Kercher was moving in. They said she was quiet and nice—from outside London. They urged me to come back soon so we could "get the party started."

I couldn't wait to return. But I'd also been chastened by my first trip to Perugia. A few days after Deanna and I got to Germany, I broke out with a gigantic cold sore on my top lip that Dolly and I figured must be oral herpes—from Cristiano. To my great embarrassment, Dolly had to take me to the pharmacy to find out how to treat it. I couldn't believe this was the first wild thing I'd done in my entire life and—bam! I'd made an impulsive decision, and now I'd have to pay a lifelong consequence.

I was bummed knowing I'd have to take medication forever. Even more humiliating was that from here on out I'd have to explain to potential partners that I might be a risk. I gave myself a hard time, but after a few days and a lot of conversations with myself, I settled down. I vowed I'd be more careful in the future. After my luck had changed in finding the villa, I'd had a stroke of bad. I told myself if this was the worst thing that happened, I could deal with it.

Chapter 3

September 2007

Perugia, Italy

I met Meredith on September 20, 2007, the day I moved into
No. 7, Via della Pergola. Half Indian, she was exotically
beautiful, a Brit majoring in European studies. In the month
she'd been in Perugia, she had already become part of a close-
knit group of British girls. As she stood in my doorway chatting while I unpacked, I understood how that had happened so
fast. She was friendly and game. "Come out with me tonight and
meet my friends," she said. "I'll introduce you to all the people
and places I've gotten to know in Perugia. You're going to love
it here."

The newest and youngest of four roommates, Meredith and
I had a lot in common. We were both children of middle-class,
divorced parents, and, at twenty-one, she was just a year ahead
of me in school. We'd each pushed ourselves hard to make this
year in Italy happen. Now that I was finally here, the hours and
hours I'd worked in Seattle—early in the morning as a barista,
late into the night for a local catering company, and training a

girls' soccer team in between—seemed unquestionably worthwhile. Meredith, a longtime Italophile, had been crushed when her British university turned her down for the program abroad, but she fought the decision and won. Maybe that's why we each brought once-in-a-lifetime determination to our experience.

Talking to her that first day, I was shocked to find how truly little I knew about my new city. I assumed Meredith and I would be classmates at the University for Foreigners, but she was enrolled at the University of Perugia. I couldn't believe I'd been so laser-focused on my own program that I'd overlooked a local college, with a whopping thirty-four thousand students, just a ten-minute walk from our villa. I could have found out all about it if I'd only bothered to do a search on Google. But I had such a brochure image stuck in my head of Perugia as a tranquil, almost monastic place. As it turned out, more than a quarter of the city's population were students, and while Perugia is more than two millennia older and way more picturesque, it's a college town much like Ann Arbor or Berkeley or Chapel Hill.

It was my first day here and reality had already punctured my expectations.

At dinner, I discovered that Meredith's friends fit the reserved British profile. I'm sure I struck them as a stereotypically loud American. I was energetic and outspoken, even by nonconformist Seattle's standards, and I was probably louder than I meant to be. While we were sitting around the restaurant table sipping wine and eating pizza, I started singing some song that was popular then. But what drew laughs in Seattle got embarrassed looks in Perugia. It hadn't dawned on me that the same quirks my friends at home found endearing could actually offend people who were less accepting of differences. A person more attuned to social norms would probably have realized that immature antics didn't play well here.

So I was glad I could hang out with Laura, Filomena, and Meredith at home. Even though Meredith was definitely more mainstream and demure than I'd ever be, and Laura and Filomena were older and more sophisticated, I felt comfortable in their company. They seemed to accept me for me right from the start.

During my first month in Perugia I spent more time with Meredith than anyone else. I liked her a lot, and she seemed to enjoy being with me. I could already see us keeping in touch by e-mail when our year abroad was over. Maybe we'd even end up visiting each other in our hometowns.

The University of Perugia started earlier than the University for Foreigners, so Meredith was in class on my first full day living in Perugia. I went exploring alone, stopping at the only place that was familiar—a small café where Deanna and I had eaten all our meals in late August. I decided to go and say hi to the tall, balding, good-humored barista who'd figured out our ridiculous espresso + milk + chocolate mixture. I couldn't remember his name, but when you arrive knowing no one, you're grateful for any friendly acquaintance. It turned out, however, that he'd moved on. His replacement was an athletic guy about my age named Mirko. He had black hair, blue eyes, and a huge grin. I told him I was new in town, a student. He said he was more into work than study. By the time I left for home that day I had the slimmest inklings of a crush.

During the lull before my semester began I dropped by the café several afternoons for a *caffè macchiato* or a glass of white wine and a little flirting. After my oral-cold-sore-inducing make-out session with Cristiano, this was sweet and innocent.

My new favorite pastime was the old Italian custom of long, relaxed lunches at home. Meredith and I ate with Laura and Filomena, who changed from skirt suits into cut-offs and flipped on the TV. Their soap opera was just background noise for me. I'm not a TV person at all, much less a soap opera fan, but I thought it was funny that the dialogue in theirs sounded exactly the same as any American soap. Understanding about one word in five, I could still follow what was happening. *You slept with WHO?! Let's run away together!* Soap operas, I learned, are another universal language.

When Filomena and Laura went back to work, Meredith and I would sunbathe on the terrace and talk. She read mysteries. I was teaching myself to play Beatles songs on the guitar. One day she said it reminded her of when she and her older sister used to turn their CD player way up and sing along.

I loved our easy togetherness. We told each other about life back home and what we were thinking about doing after we graduated. She said she might want to become a journalist, like her dad. She lent me clothes—her feminine look was more in keeping with Perugia than my old jeans and boyish T-shirts—and she helped me with my Italian grammar. Before beginning classes at the University of Perugia, Meredith had taken a crash course at the University for Foreigners, to brush up on her language skills. I introduced her to new music and listened to her stories about her family, especially her mom, whose bad health worried Meredith so much that she didn't go to the corner *bottega* without her British cell phone in her pocket.

I told her about my parents and stepparents—that Dad had been with Cassandra since I was little, and Mom had met her husband, Chris, when I was ten.

And we did what all girls do: we talked about the guys we liked in Perugia and the ones we'd left behind. I told her about my growing crush on Mirko and about my ex-boyfriend from Seattle, DJ. "DJ and I were together for eight months," I said. "We broke up because I was coming here and he was going to China for the year. We're still friends, though."

"What's he like?" Meredith asked.

"Completely eccentric," I said. "He has a Mohawk, wears this shabby red kilt, and goes everywhere barefoot, except when he goes climbing. Then, I promise, he wears shorts and shoes."

Meredith laughed. "He sounds like your type. Do you think you'll get back together?"

"I can't tell," I said. "What about you? Do you have a boyfriend?"

"I dated a guy pretty seriously," Meredith said. "We were together for a few months. I have real feelings for him, but I'm too young to get serious. I still have two more years of university. We broke up right before I came to Italy."

"Well, you don't want to feel weighed down by decisions while you're figuring life out."

We encouraged ourselves with affirming smiles and cheap local red wine.

Meredith and I did a lot of routine things together—like walking to the grocery store and going to the rental office. She asked me to snap photos of her standing in front of the picture window in her bedroom. "I want my family to see my view," she said.

One afternoon, when I discovered a vintage clothing shop downtown, I was so excited that I went home and immediately brought Meredith back with me, the sort of thing I'd usually do with Brett or Madison.

"These clothes are definitely more offbeat than I'm used to," Meredith said, "but they're awesome." She tried on a few things, coming out of the dressing room to model each one and discuss all the places she could wear it. She bought a sparkly silver vintage dress she said she'd wear for New Year's Eve in London.

It made sense that Meredith and I were closer to each other than to our other flatmates—we were both trying to learn a city and a language we didn't know. Filomena and Laura were longtime friends, older, finished with college, and Italian. To them, Perugia was the same old, same old.

While I waited for the semester to start, I tried to read in Italian and tested new vocabulary wherever I could. One day, I went to the Coop, a supermarket in Piazza Matteotti, gathered my groceries, and went to the register to pay. *"Busta?"* the cashier asked me.

I didn't know the word. Was it *envelope*? Was she asking me if I wanted to buy envelopes? I could feel the people behind me in line shuffling impatiently. I was about to respond no, but she read my confused expression before I got the word out. She shook a plastic shopping bag in my face. *"Busta?"*

I reddened. *"Sì, sì, busta. Grazie. Scusa,"* I said.

I knew I shouldn't have been embarrassed, but I didn't want to be regarded as a tourist. I didn't want attention brought to my ignorance of the language.

I didn't let my mistakes keep me from getting to know my neighborhood or my neighbors a little better. Each time I went to the Internet café to Skype with DJ or chat online with Mom, I'd talk to the guy who ran it, Spyros, a Greek in his late twenties. We talked about the same things that filled my conversations with my UW friends—mainly our ideas and insecurities. He graciously welcomed my sputtering attempts to speak in Italian about more than the weather. This was a little different from home, where Laura and Filomena found my deficient Italian entertaining and chuckled at my slipups.

A few times a week I hung around the coffee shop chatting with Mirko about what we each liked to do and about our personalities. Me: serious, goofy early bird. Him: playful, easy-going night owl.

One afternoon I asked, "Do you know where I can go hear live music?"

"No, I like sports," he said. "Do you like Inter?"—Milan's popular professional soccer team.

"I prefer to play soccer than watch it. I was a defender on a premier team," I said. I could tell he was picturing me in a kid's church league.

"Are you good?" he asked.

"Do you know the American expression 'sly like a fox'?" I asked. "That's what I was—fast and reliable for finding an opening and stealing the ball. My teammates nicknamed me Foxy Knoxy."

The next time I went to the café, I watched him work, looking for signs of a connection between us. There was music playing in the background—popular dance music from a local radio station. "Do you like music?" I asked.

"I like to dance," he replied. "Do you?"

Is he a lowbrow party guy? I wondered. That's not what I was hoping for. "I'd rather sing and play guitar," I answered.

Maybe we had reached an impasse. Just then, Mirko said, "I thought of a place you'd really like—for pizza."

"Let's go sometime," I said. I thought, *I can't believe I just asked him out.*

"How about today?" he asked. "I get off at five."

I was excited. Mirko was nice, laid back, and interested in me.

When I arrived back at the café to meet him, he was just taking off his apron. We walked down Corso Vannucci, Perugia's main commercial street, and turned onto a quieter side street of shops and restaurants. People were lined up outside the pizzeria, waiting for a table.

"Do you want to eat at my place?" Mirko asked. "We can watch a movie."

"Sure," I said, and instantly felt an inner jolt. It came from the sudden certainty that we would have sex, that that's where our flirtation had been heading all along.

We carried our pizza boxes through Piazza Grimana, by the University for Foreigners, and down an unfamiliar street, past a park. Mirko's house was at the end of a gravel drive. "I live here with my sister," he told me.

During dinner at his kitchen table my thoughts battled. Was I ready to speed ahead with sex like this? I still regretted Cristiano. But I'd also been thinking about what Brett and my friends at UW had said. I could picture them rolling their eyes and saying, "Hellooo, Amanda. Sex is normal."

Casual sex was, for my generation, simply what you did.

I didn't feel that my attitude toward sex made me different from anyone else in my villa. I knew Meredith hadn't been with anyone since her serious boyfriend in England. Filomena had a steady boyfriend, Marco Z., in Perugia. And while Laura was dating and sleeping with a guy she thought was sweet but clingy, she encouraged sex outside relationships.

From the start, all four of us were open to talking about sex and relationships. Laura insisted that Meredith and I should just have fun. Filomena was a little more buttoned-up. She couldn't understand how, with our history together, DJ and I could just be friends and inform each other about our romantic exploits over Skype.

I considered Mirko across the checkerboard table as he devoured his pizza. He was part of the small circle of familiar faces I'd started to create for myself in Perugia.

We didn't talk much at dinner. I dawdled, asking him the standard questions about himself. He dodged them and asked, "What movies do you like?"

"I like anything that's not scary," I said. "I'd really like to see a classic Italian film."

"I have a funny one on DVD," he said.

Of course the TV was in his bedroom.

"It's a little cold. Let's get under the covers," he coaxed.

I did, fully clothed except for my sneakers.

The movie was so juvenile I could barely pay attention. I was mostly focused on how the night would unfold. I liked Mirko, but I didn't know him. He was attractive, and his confidence was charming. His taste in movies was bad, though. Still, I told myself, *People have flings.*

When the movie ended, Mirko clicked off the TV. Without speaking, he leaned over and kissed me. I kissed him back. It was happening.

As soon as it was over I quietly got back into my clothes, wondering what I thought of my newfound freedom. I was proud of myself for having a no-strings-attached consensual encounter, but I felt awkward and out of place. I didn't yet know if I'd regret it. (Nor could I anticipate that my private, uncertain experiment would become my public undoing.) "I'm sorry," he said, "but you have to go now. My sister will be home soon. I'll walk you to the University for Foreigners. You can find your way from there."

We didn't talk as we walked past the park. When we reached

the university, he kissed me good-bye on both cheeks. The standard Italian hello and good-bye among casual friends was as unromantic as a handshake would have been in America. "We should do that again sometime," he said. I nodded, perplexed by the disparate emotions bouncing around in my head.

I walked back to the villa alone, feeling both exhilarated and defeated.

The next morning, I told my roommates I'd had sex with Mirko. "I feel conflicted," I said. "It was fun, but it was weird to feel so disconnected from each other. Is that just me?"

Laura absolved me. "You're young and free-spirited. Don't worry about it."

That made me feel a little better.

A few days later, I stopped by the café, and Mirko invited me to his place again. I shoved my ambivalence aside and agreed. As we walked from the café, he smiled at me and asked me how school was going. "Fine," I said. "How's work?"

"Pretty slow, now that the tourist season is over."

We didn't hold hands.

I followed Mirko down the gravel drive and into his house. I wanted to turn around and run, but somehow I couldn't. I found myself inside his bedroom. Mirko playfully pushed me on to his queen-size bed, but when he put his hand down my jeans I balked. "I have to go," I said. I didn't say why. I just threw on my shirt and left, walking alone up the road, past the park, past the University for Foreigners, home. I didn't feel free or sophisticated. I felt a twinge of regret.

I was too ashamed and embarrassed to go back to the café after that. Was there something wrong with me? Or was it with him? Either way, I couldn't bear to run into him again.

I was alone with Meredith when I told her about fleeing from Mirko.

"I feel like an idiot."

"Amanda," she said, consolingly, "maybe uninvolved sex just isn't for you."

———————

Months later Meredith's friends, our roommates, and especially the prosecutor would say that Meredith's and my relationship had soured—that we had fought over men, my manners, money. This wasn't true. We never argued about anything. We were just getting to know each other, and I thought we'd developed a comfortable familiarity in a short time—a process that probably moved faster because everything around us was new and unfamiliar. We shared a house, meals, a bathroom. I treated Meredith as my confidante. Meredith treated me with respect and a sense of humor.

The only awkward interaction we had was when Meredith gently explained the limitations of Italian plumbing.

Her face a little strained with embarrassment, she approached me in my room and said, "Amanda, I'm sorry to bring this up with you. I don't know if you've noticed, but with our toilets, you really need to use the brush every time."

I was mortified. I knew that Meredith was uncomfortable saying it. I would have been, too. I said, "Oh my God, I'm so sorry. I will totally check and make sure I leave it clean."

We laughed a little nervously. We didn't want to hurt each other's feelings.

For two weeks in mid-October, tents and tables filled all the squares around Corso Vannucci for the annual Eurochocolate festival. The smell of chocolate around town was inescapable. Laura told me about the chocolate sculpture carving. It was done in the early mornings, so the next day, I went to Piazza IV Novembre to watch. The artists started with a refrigerator-size block of chocolate. As the chiseled pieces flew, assistants gathered chips and shavings into small plastic bags and threw them to the rowdy crowd. When a chunk of chocolate with the heft of an unabridged dictionary fell, onlookers screamed and reached across the barrier. I was shorter than most of the people around me, but I jumped up and down, yelling, *"Mi, mi, mi!"*

I was amazed when a worker plopped the chocolate in my arms. People reached at it, picking little pieces off as I disentangled myself from the crowd. I rushed home, trying to get there before the block melted on my shirt. I unloaded it on the table and said, *"Voilà!"* Later, Meredith and I made chocolate chip cookies out of part of my winnings, trying to recreate the Toll House recipe from guesswork and memory.

Another afternoon, I returned to the festival with Meredith. I flipped the video switch on my camera and acted like a TV journalist. "Tell me, Meredith, what do you think about being here at the Eurochocolate festival?"

Meredith laughed and said, "No, no, don't film me." She pushed the camera away. She didn't like being the center of attention.

Neither of us felt we had to go far to be entertained. More

days than not, three of the four guys who lived downstairs—Giacomo, Stefano, Marco M.—and another friend, Giorgio, dropped in during lunch and again after dinner for a stovetop espresso and, almost always, a joint. A few years older than Meredith and me, they were each equal parts big brother and shameless flirt. Students at the University of Perugia from Marche, the countryside east of Perugia, they'd sit around, get high, and gab about soap operas and game shows, movies, music—nothing in particular. Of the four, Giacomo, tall and sturdy like an American football player, with pierced ears, buzzed hair, and doe eyes, was the quietest and shiest. He played the bass, studied Spanish, and spoke English better than his roommates. When the guys weren't at home, upstairs at our apartment, or at school, they were often playing basketball on the court in Piazza Grimana. One day, hoping to replicate the football games I loved to play with my guy friends at UW, I asked if I could come shoot hoops with them. "Sure," they said. But when I got there I realized they thought I meant only to watch them play. It was another way in which my Seattle upbringing had left me unprepared for the cultural strictures of my new environment.

Around our house, marijuana was as common as pasta. I never purchased it myself, but we all chipped in. For me, it was purely social, not something I'd ever do alone. I didn't even know how to roll a joint and once spent an entire evening trying. I'd seen it done plenty of times in both Seattle and Perugia, but it was trickier than I thought it would be. Laura babysat my efforts, giving me pointers as I measured out the tobacco and pot and tried rolling the mixture into a smokable package. I never got it right that night, but I won a round of applause for trying. Either

Filomena or Laura took a picture of me posing with it between my index and middle finger, as if it were a cigarette, and I a pouty 1950s pinup.

I was being goofy, but this caricature of me as a sexpot would soon take hold around the world.

Chapter 4

October 2007

My big lesson on the first day of school at the University for Foreigners had nothing to do with academics. I'd shown up early for my 9 A.M. grammar class and then waited, alone, checking and rechecking my schedule, wondering if somehow I'd come to the wrong place. Finally, just when I was about to give up, everyone arrived, en masse, at 9:15. That's when I learned that Italian time means *T* plus fifteen minutes, a tidbit I thought would probably be more useful to me than any verb I might learn to conjugate. I was even more charmed and appreciative the morning my pronunciation teacher announced that she needed a cigarette and suspended class while all fifteen of us walked outside for a quick shot of espresso or a smoke. Italians, I was coming to see, embrace any chance to have a good time. It was something I wasn't used to.

I went to school for two hours, five days a week. Besides grammar and pronunciation, I had a third class, in Italian culture. We all went home for lunch at noon, and I spent the rest of the day and night doing whatever I wanted. My teachers didn't give homework, so I'd sit on the terrace or, when the days cooled, at my desk with a grammar book and a dictionary, making my way, one word at a time, through the Italian translation of *Harry Potter and the Chamber of Secrets*.

A lot of people would have traded places with me in a nanosecond. I was living in Italy, young and unfettered. But I quickly found that, for me, all the freedom came with a downside: the empty hours made me feel irresponsible, and I knew I needed somehow to fill them.

When I told my roommates in early October that I wanted a job, Laura arranged a meeting for me with a friend of hers. After lunch one day, she walked me over to Piazza Grimana, by the University for Foreigners, and introduced me to a scruffy young guy. "Juve, this is Amanda. Amanda, this is Juve. Good luck!" she said, turning to walk away.

Juve smiled and shook my hand. He spoke perfect, albeit stilted, English. "So you are looking for a job?" he said.

"Yes, I have experience as a barista."

"You are American?" he asked.

"Yes, I'm from Seattle."

He put his arm around my shoulders and started walking me down the street, away from the university. I felt awkward, because I was looking for a job, not a boyfriend. "My boss, Diya Lumumba—people call him Patrick—owns a new bar, Le Chic," Juve said. "It's a good place, small, and we are building up our clientele. I give out flyers during the day and bring in customers in the evening. I make sure business happens."

"Is that where we're going?" I asked.

The answer was no.

Instead we walked to Juve's apartment, where he made us espresso and we took turns playing his guitar. It was the weirdest job interview I'd ever had. I wasn't sure what to say or do. Was I supposed to talk about my work experience pulling coffee at a café and working for a Seattle caterer? Or were we two coworkers just hanging out? I'd expected him to ask me questions, but I had the feeling I'd already gotten the job on our handshake. Like so many other experiences in Perugia, this one made me feel off-balance.

"About the job," Juve finally said. "It is straightforward and easy. I will give you flyers to hand out in school. Invite your classmates to come to Le Chic. Keep asking them. Around nine P.M., we will meet at the bar, Patrick will open the doors, and we will help him get ready. Then we will go to Corso Vannucci and hand out more flyers and direct people to Le Chic. When we have no flyers left, we will help Patrick get drinks, stock snacks, and make sure people are having a good time. When customers leave, we bring in more."

I was hired to work at the bar from 9 P.M. to 1 A.M., making €5.00—about $7.25—an hour. "Handing out flyers doesn't count as work time," Juve said.

"Okay," I answered. "So, what's next?"

"You need to meet Patrick. I will tell him I know you and will teach you how to do the job."

I met Patrick at lunchtime the next day, at the university snack bar. He was originally Congolese and spoke Italian but no English. "Do you understand what I'm saying?" he asked.

"Pretty well," I said.

Like Juve, Patrick wasn't interested in my work experience. Looking back now, I'm sure they hired me because they thought I'd attract men to the bar. But I was too naïve back then to get that. I still thought of myself as a quirky girl struggling to figure out who I'd be when I grew up. I now realize that the point of the job "interview" was to see if my looks were a draw or a liability.

Patrick said, "Sometimes the job is serving, sometimes cleaning, sometimes being friendly and welcoming."

"I'm outgoing," I said, still trying to sell myself as a hard worker. "I love talking to people."

"Good. Juve will train you, and I will see you tonight!" Patrick stood up and kissed me on each cheek. Juve handed me a stack of flyers. "The students are leaving classes," he said. "Hand these out. And congratulations!"

Just before nine o'clock I went to Juve's apartment, and we continued on to Le Chic together. The wooden door was open, revealing a dim, tiny vestibule with a bar and a seating area beyond. The walls were brick, and the whole place was dark. It looked like a cellar.

Patrick was standing behind the bar. "Welcome," he said, as he handed me a menu and started pointing out which beers were on tap, which in bottles, and the different liquors for cocktails and specialty drinks. "Would you like something?"

"I'm good for now, thanks," I said, translating the American idiom directly into Italian. Patrick looked at me blankly. "No, thank you," I said, correcting myself.

It turned out that working at Le Chic wasn't nearly as easy or as straightforward as billed. In fact, it was bewildering. I didn't always understand Patrick's directions, especially over the pulsating music, and I had to rely on Juve to translate. It was hard to keep track of orders let alone customers, whom I could hardly expect to stand in one place all night. I couldn't maneuver trays and had to hand-carry the full glasses two at a time. It was my job to make sure customers kept drinking, and I had to watch carefully so I could pounce on not-quite-finished cocktails and replace them before the last swallow. It felt like a lot to juggle, even without the added challenge of trying to stay awake until 1 A.M. on school nights.

"Have a good time" was my main instruction from Patrick. If I was having fun, the customers would, too. This wasn't at all what I had thought I was getting myself into. As a barista at home, I was friendly with the regulars, but I wore an apron and stayed behind the counter, protected. At Le Chic, I liked being out and around people, but the job made me feel used and unsure of myself. Still, once I commit to something, it's hard for me to admit that it's not working out.

Patrick always offered me drinks on the job, and I couldn't figure out what kind of message he was sending me. Since I wasn't a big drinker I'd either turn him down or nurse a single glass of white wine all night.

Every day, Juve met me outside my grammar class with a new stack of flyers, and I'd hand out a few between and after my classes. I dreaded the hour between 9 P.M. and 10 P.M., when I'd have to stand by myself in Perugia's main square, Piazza IV Novembre, calling out, "Le Chic. Via Alessi. Le Chic. Via Alessi." I felt vulnerable.

Piazza IV Novembre, home to both the Duomo, a massive fifteenth-century Gothic cathedral, and an elaborately carved pink-and-white marble fountain, was the town's main meeting spot. At night it filled with loud students milling around drinking beer from plastic cups. It reminded me unhappily of the fraternity bashes I'd attended as a freshman at UW. I'd gone to those parties, danced with those people, drunk too much. It took me less than a semester to figure out how much I disliked it. Being in school in Perugia, I felt as if I'd circled back to the same spot—ironic, since I'd come to Italy to figure out how to be my own person.

My job made me feel like a bull's-eye in the middle of the chaos. Guys continually came up to me to flirt, saying they'd stop by Le Chic only if I promised to be there. Brushing them off, as I would have liked, would have been bad for business. So I hoped my chirpy "You should come by" came off as inviting for Patrick's sake and not too suggestive for mine.

It was confusing to me. I was open to new people and experiences, but I kept ending up in situations I didn't want to be in. Working for Patrick and Juve was part of that.

Since most of my days included standing there mute with my arm outstretched to passersby who didn't acknowledge that I was at the other end of the four-inch-by-five-inch sheet of colored paper, I was always relieved when my stack of flyers dwindled, and I could leave.

But no matter how many flyers I gave out, Le Chic wasn't catching on. Meredith came to visit me there a few times so I wouldn't be bored or alone, and once, she brought her girlfriends. But I could see why they didn't come back. Le Chic didn't get a lot of foot traffic, so the dance floor was usually empty. The bar felt forlorn—not exactly a recipe for a good time. Patrick was jovial and did his best to make it welcoming, but it was still noisy and dark inside and attracted a crowd of older men—often friends of Patrick's—and not students.

There was nothing truly dangerous about Le Chic, but its seediness did hint at Perugia's dark side. What I didn't know when I arrived was that the city had the highest concentration of heroin addicts in Italy. I never heard about the high level of trafficking and drug use until I was in prison, bunking with drug dealers. During my trial, the prosecution and the media seemed to take for granted that our neighborhood was bad and our little villa a deathtrap.

Even without knowing this, my mom worried about my safety—a lot. One day, while I was e-mailing back and forth with her at the Internet café, she asked, "Who should I call if I can't reach you?"

"We don't have a home phone, but I can give you Laura's number," I wrote. "But honestly, Mom, I think I'm safer here than in Seattle. My friend Juve walks me home from work most nights, and Perugia is much smaller than Seattle. I've really made a lot of friends."

"Okay," Mom wrote back. "I feel better."

I believed what I said—not because I had reason to but because I was in love with the city's many charms. And I didn't pick up on some obvious clues.

One night, when Le Chic was closing and Juve couldn't walk me home, I saw an acquaintance of Meredith's. I didn't know his real name, only that Meredith and her girlfriends had nicknamed him Shaky because of the way he danced. He offered me a ride home on his scooter. I figured a friend of a friend was close enough to trust. I figured wrong.

We whizzed through the narrow streets. As we approached the intersection where the villa was, he slowed and yelled over his shoulder, "Would you like a cupcake? I know the best bakery in Perugia, and it's open all night."

"No. I'm tired. I just want to go home."

"Come on. It's nearby."

"No, thanks," I said, just as we passed my house.

We went to the bakery, where I refused to get anything. "I don't even like cupcakes," I said. "Now home."

"My home," he said.

"No!" I glared.

"Just for a minute. I have to pick something up."

"Okay," I said, feeling that it was not okay at all. But I had no idea where I was and no other way to get where I was going.

Shaky's apartment was tiny and cramped with people. He took me to his bedroom to wait while he went off to do something. After a few minutes, he came back with a beer for me.

I said, "If you don't take me home right now I'm going to walk." Luckily for me, since it was an empty threat, he shrugged, turned around, and we left. When we got to my driveway I climbed off the scooter without saying good-bye and stormed inside.

I was angry, and bursting to tell Meredith. She sighed. "I'm so sorry," she said. "He tried to do the same thing with my friend Sophie. But he was so responsible the night our friend was sick, I still really trust him."

After that, Meredith came up with a plan. She always went out with a group of girlfriends, so she felt protected by the pack. But knowing I was often on my own, she said, "If you come back to the villa at night and I'm not here, make sure to text me to say you've gotten home safely."

It was comforting to know that if she didn't hear from me she'd realize something was wrong and would get help.

One night when the bar was slow, Patrick decided to close early. I texted Meredith, who said she'd meet me at the fountain by the Duomo, three minutes away.

As I made my way through the mass of drunk students in Piazza IV Novembre, I saw two of our downstairs neighbors, Giacomo and Marco. Giacomo handed me a beer, and I pushed my way through the crowd to find Meredith. When we had rejoined the guys, they introduced us to a friend who, I'd later learn, had moved to Italy as a kid, from Ivory Coast. His name was Rudy. They sometimes played pickup basketball with him.

The five of us stood around for a few minutes before walking home together. The guys invited us to their apartment, but Meredith and I first stopped at ours to drop off our purses.

"Ready to go downstairs?" I asked her.

"You go. I'll be down in a second," she said.

When I opened the door to the downstairs apartment, Giacomo, Marco, Stefano, and Rudy were sitting around the table laughing. "What's funny?" I asked.

"Nothing," they said sheepishly.

I didn't think another thing about it until months and months later, when it came out in court that just before I'd opened the door, Rudy had asked the guys if I was available.

A short time later, Meredith came in and sat down next to me at the table. The guys passed us the joint they were smoking. We each inhaled, handed it back, and sat there for a few minutes while they joked around in Italian. Tired and a little stoned, I couldn't keep up with their conversation. After a little while I told Meredith, "I'm going up to bed."

One day in mid-October, about three weeks after I arrived, Meredith and I were walking down Via Pinturicchio to try out a new grocery store that was supposed to be cheaper than the Coop we usually went to downtown. I didn't know it then, but it was just a few doors down from Perugia's courthouse.

"Have you met any guys you like yet?" I asked Meredith.

"Giacomo," she said, shyly but decisively. She had talked about our downstairs neighbor before. "I think he's cute and nice."

Not many nights later, the guys invited all of us in the house on an excursion to Red Zone, a popular club just outside of town. I was excited. It wasn't usually my scene, but I'd decided to try something different and had already been to two downtown dance clubs, Domus and Blue Velvet. To my surprise, I'd had a decent time.

Laura and Filomena stayed home, but Meredith invited her friend Amy to come. The guys brought a friend from Rome named Bobby, whom I'd met once before. I had a cold sore then and was so self-conscious about it I just wanted to hide. Bobby said, charmingly, in English, "Why does it bother you? Many people get cold sores."

Red Zone took up an entire warehouse. It was the largest, most over-the-top dance club I'd ever been to. The line to get in snaked around the building, and people were crammed in as if the place had been vacuum-sealed. It was hard to find any air. Bright lights flashed red, green, and blue, and the heavy bass seemed to travel through the cement floor and into my bones. Somehow we snagged a table, and Stefano ordered a round of sweet, electric blue drinks. I don't know what was in them, but I got drunk almost immediately. We were listening to the music and laughing, getting up to dance every now and then. It must have been 102 degrees, and I was sweating, dripping. Bobby tried to talk to me, yelling over the music.

When I went to the bathroom, he followed me and waited outside the door. As I stumbled out, I grabbed onto him and kissed him on the mouth.

"Do you like me?" Bobby asked.

I nodded.

Then he kissed me back.

Just then, Marco passed by and started whooping and congratulating Bobby on our hookup. I have no idea how long we stayed at the club. When it was time to go, Stefano went for the car, and Bobby and I stood on the curb outside, kissing. Giacomo and Meredith stood slightly apart from us, entwined.

When we got home, Bobby followed me to the front door.

"Do you want to come in?" I asked.

"Are you sure?"

I nodded. This was the first time I'd invited a guy into my bed since I'd arrived in Perugia. We went to my room and had sex. Then we both passed out.

The next morning I got up before he did, got dressed, and went to make myself breakfast. Bobby came into the kitchen a few minutes later.

We were eating cookies when Laura came out of her bedroom. I'd never entertained a lover at the villa for breakfast, and it was awkward, despite Laura's proclaimed sense of easy sexuality. All three of us tried to ignore the feeling away.

After breakfast Bobby left to return to Rome. I walked him to the door. He smiled, waved, and walked away.

I didn't feel the same regret I'd had after sex with Mirko, but I still felt the same emptiness. I had no way of knowing what a big price I would end up paying for these liaisons.

A few minutes later, Meredith came upstairs. She and Giacomo had slept together for the first time, and she was giddy. It had been a wild night at No. 7, Via della Pergola, but it turned out to be a one-time thing.

A couple of days later Juve told me that Patrick wasn't entirely happy with my job performance and wanted to meet with me on the Duomo steps. I knew I had been slow delivering cocktails and that I wasn't attracting customers as they'd hoped.

To my surprise, Patrick was kind. "You really need time to

pick up waitressing," he said. "On busy nights I need someone who's more experienced. You can keep working the slow
nights—Tuesdays and Thursdays—if you'd like. That way you'll
be learning."

I was relieved. I liked the purposeful feeling I got from working, but I knew this wasn't the right job for me. I'd already started
leaving my name at bookstores and other places around town.

I was just beginning my second month in Perugia, and I still
felt uprooted from Seattle. But I felt I was finally starting to hit
my groove. I recognized the faces I passed every day as I walked
to and from school. More important, I felt that the choices I'd
made were educating me. I just had to wait for what, and who,
would come next.

Chapter 5

October 25–November 1, 2007

B y chance.
I found my roommates by chance. I saw a poster advertising a performance of a string and piano quintet by chance. I met Raffaele Sollecito by chance.

On Thursday, October 25, Meredith and I went together to the University for Foreigners to hear Quintetto Bottesini. We sat together by the door of the high-ceilinged hall. During the first piece—Ástor Piazzolla's "Le Grand Tango"—I'd just turned to Meredith to comment on the music when I noticed two guys standing near us. One was trim and pale with short, disheveled brown hair and frameless glasses. I was instantly charmed by his unassuming manner. I smiled. He smiled back.

When Meredith left at intermission to meet friends for dinner, the guy walked up to me.

"Are these seats open?" he asked in Italian.

"Yes, please to sit," I said in my imperfect Italian.

"I'm Raffaele," he said, switching to English.

"Amanda."

Later I would wonder what would have been different if this

hadn't happened. What if Meredith had stayed at the concert? What if Raffaele had gotten there in time to get a seat? Would we have noticed each other? Would he, naturally shy, have introduced himself without the excuse of a needed chair? Would never knowing him have changed how I was perceived? Would that have made the next four years unfold differently? For me, maybe. For Raffaele, absolutely.

But we did meet. And I did like him. Raffaele was a humble, thoughtful, respectful person, and he came along at the moment that I needed a tether. Timing was the second ingredient that made our relationship take off. Had it been later in the year, after I'd found my bearings and made friends, would I have needed the comfort he offered?

Waiting for the return of the quintet, we talked. His English was better than my Italian.

"Do you like the performance?" he asked.

"Yes, I love classical music," I said.

"That's unusual for someone our age," he replied.

He was right. The rest of the audience looked three times older than we were, and I hadn't yet found anyone my age in Perugia who talked about classical music. Grabbing a friend and going to Benaroya Hall to hear the Seattle Symphony on my UW student discount was something I had done as often as I could back home.

During the second half, I whispered to Raffaele just as I had to Meredith. It was nice to have a shared, uncommon interest.

When we stood up to leave, he asked for my number. In Perugia, where I'd gotten this question a lot, my stock answer was no. But I thought Raffaele was nerdy and adorable—definitely my type. He was wearing jeans and sneakers that evening. Like

DJ, he had a pocketknife hooked to his belt loop. I liked his thick eyebrows, soft eyes, high cheekbones. He seemed less sure of himself than the other Italian men I'd met. I said, "I'll be working later at Le Chic on Via Alessi. You should come by."

At 10 P.M., when I got to the bar, a handful of customers were drinking beer. Juve pumped up the music, and I tried to keep busy doing mindless tasks—refilling the snack bowls, wiping tables, ensuring the bathrooms were clean, checking my appearance in the mirror. I was grateful for the distraction while I waited to see if Raffaele would show up.

Every time I heard the door open, I looked up hoping it was Raffaele. When he walked in with three friends, my stomach did a nervous flip. I went over to their high-top table with menus and a huge grin. I found out later that, to get his friends to come, Raffaele had promised to buy their drinks.

For the next hour, I waited on other customers but only really paid attention to one. It was the first time in my lackluster waitressing career that I did exactly what Patrick had asked for all along: I magically materialized at Raffaele's table well ahead of his last sip.

"Another round?" I asked.

"No, thanks," Raffaele said. "When do you get off?"

"In about half an hour," I said. "Would you like to walk me home?" Walking with a guy was a tactic I'd used a few times in Seattle to figure out if I wanted to see him again. A walk is a much smaller commitment than a date.

We wandered slowly through town, away from my villa, to the far side of Corso Vannucci—Piazza Italia. We stopped at an overlook in front of a low brick wall.

We stood high above the Tiber Valley, staring out at the speckled lights below. "This is the perfect place to think," Raffaele said. His nervousness was palpable—and contagious. A fidgety silence hung between us as we gazed out, until, gradually, we looked less at the view and turned toward each other.

It wasn't an electric first kiss that bound us together. It was gentle and soft—comfortable and reassuring.

I don't know how long we stood with our arms wrapped around each other. When we pulled apart, the air was so sharp I could see my breath. But I knew that this was the warmest, safest, most enveloped I'd felt since August, when I'd hugged good-bye the people I loved most. After a month on my own, the exhilarating feeling of taking charge of my life had receded a bit. I wavered between feeling self-confident and needy. I was reveling in everything new and feeling homesick for the familiar. Raffaele, with a single kiss, had bridged the gap. He was a soothing presence.

Afterward we walked past the fountain in Piazza IV Novembre. Another five minutes and we'd be at my house. I so badly wanted to extend the moment. "Do you like marijuana?" I blurted.

"It is my vice," Raffaele said.

"It's my vice, too," I said. I loved the phrase in Italian.

Raffaele looked surprised, then pleased. "Do you want to come to my apartment and smoke a joint?"

I hesitated. He was basically a stranger, but I trusted him. I saw him as a gentle, modest person. I felt safe. "I'd love to," I said.

Raffaele lived alone in an immaculate one-room apartment. I sat on his neatly made bed while he sat at his desk rolling a joint. A minute later he swiveled around in his chair and held it out to me.

We talked as we smoked. He was twenty-three, from Bari, in southern Italy, and three weeks away from getting his degree in computer science. "I'm moving to Milan in the new year," he said. "I'm hoping to get a job designing video games."

We learned we had a third language in common, German. When I told him I'd studied Japanese in high school, he said he loved *Sailor Moon*, a Japanese comic book about girls with magic powers fighting evil; the TV series it spawned had been my favorite when I was younger. I was surprised by how childish his comic book interest was, but I thought his willingness to admit it was endearing.

The marijuana was starting to kick in. "You know what makes me laugh?" I asked. "Making faces. See." I crossed my eyes and puffed out my cheeks. "You try it."

"Okay." He stuck out his tongue and scrunched up his eyebrows.

I laughed.

By then, Raffaele had moved next to me on the bed. We made faces until we collided into a kiss. Then we had sex. It felt totally natural. I woke up the next morning with his arm wrapped snugly around me.

After that first night, and for seven days, Raffaele and I were a thing. We spent all the time we could together. After breakfast I'd run home to shower—his was cramped—and change for class. We'd meet up back at his apartment or mine for lunch. In the afternoons, I did my homework while he polished his thesis, which was due in two weeks, a week before his graduation. His father was planning a huge celebration at a fancy restaurant nearby.

We communicated through a hodgepodge of Italian, English,

and German—but often fell back on kisses and caresses. I loved curling up in his lap or hugging him from behind while he did the dishes. When we took a shower together, he washed my hair and then toweled me dry, even cleaning my ears with a Q-tip. To me, it was intensely tender; it felt as intimate as sex.

Meredith had just started seeing Giacomo, as a boyfriend, and she and I joked that we were living parallel lives. When we overlapped for a few minutes at home without them, we would both download. She said, "I like Giacomo, but he's shy with me when we're around other people. It really bothers me when he doesn't say hello or even acknowledge me if I run into him in town."

"Maybe you need to give him a little time," I suggested.

"Yeah, that's what I think, too," she said. "But what about Raffaele? It seems like you totally like him."

"Yeah, I really do."

And I did. He was generous with his time and with me. He had a focused attention to detail. His shirts were soft cotton and his sweaters and scarves were cashmere—all a lot nicer than my jeans and sweatshirts. And even though I didn't know anything about cars, he was proud to show me his Audi. When Raffaele found out I didn't have a signature scent, as a good Italian woman should, he took me to a fragrance shop downtown to pick one out. I'd put a drop on my arm and hold it up for him to sniff. We settled on a perfume made with sandalwood—something light and earthy that reminded me of how Perugia smelled in the morning. Raffaele paid without hesitation and handed me a pretty shopping bag tied with a blue ribbon. The experience made me feel sophisticated and, for once, truly sexy. We walked to his apartment holding hands.

That night, when we were cuddling in bed, he turned to me and said, *"Ti voglio bene"*—literally, "I wish you well." I'd heard this phrase a lot since I'd been in Perugia, and *TVB* is standard Italian text speak.

"*Anch'io ti voglio bene,*" I said—"I also wish you well." I didn't realize how much weight his three words could carry. It's what Italians say to their families, just a step below the most amorous expression, "*Ti amo*"—"I love you."

Raffaele looked at me seriously, appreciatively. "Will you be my girlfriend?"

We'd known each other for three days.

"Yes," I said, feeling a tiny twinge that I took as a warning sign. *This is moving too fast. Is Raffaele making too much of our relationship too soon?* He'd already said he wanted to introduce me to his family at graduation, and he was planning our winter weekends together in Milan. We barely knew each other.

I couldn't see how we would last, because we were a couple of months away from living in two different cities, and I was definitely going back to Seattle at the end of the next summer. Since a big part of why I'd come to Italy was to figure myself out, it occurred to me that maybe I should be alone, that I should slow things down now, before they rocketed ahead. But just because I thought it doesn't mean I did it.

It was easy to shove my doubts aside, because I really liked Raffaele. He was sensitive, and I felt calm around him. And without any solid ties, I'd been lonelier in Perugia than I'd realized.

In hindsight, I recognize that he . . . that *we* were still immature, more in love with love than with each other. We were both young for our ages, testing out what it meant to be in a caring relationship.

Being with Raffaele also taught me a big lesson about my personality that I'd tried so hard—and harmfully, in Cristiano's case—to squelch. I was beginning to own up to the fact that casual hookups like I'd had with Mirko and Bobby weren't for me. I like being able to express myself not just as a lover but in a loving relationship. Even from the minuscule perspective of a few days with Raffaele, I understood that, for me, detaching emotion from sex left me feeling more alone than not having sex at all—bereft, really.

I didn't know that this lesson had come too late to do me any good.

———

As it turned out, Halloween fell on the one Wednesday Raffaele and I were together. Unlike in the United States, kids in Italy don't go door-to-door collecting candy. Still, in a college town like Perugia, Halloween offers an irresistible excuse for students to dress up in costumes and to party—and the local bars and discos go all out to oblige them. For clubs, it is *the* number one make-money night of the year.

Patrick had asked me to show up at Le Chic even though it wasn't supposed to be my night, and Raffaele stayed home to work on his thesis. I'd been so caught up in my love life that I didn't even think to buy a costume until it was too late. So I was pretty proud of myself when I dug through my closet and found a black sweater and black pants. Raffaele helped me draw on whiskers using eyeliner, and off I went, transformed into a black cat. The bad luck superstition never occurred to me.

The town was jammed, and all the masked, wigged, mummified students made the mood in Piazza Grimana feel ominous. Of course I knew the crowd wasn't threatening, but I've always been kind of creeped out by costumes. As I passed long lines of people waiting to board buses chartered by clubs such as Red

Zone, I sent Meredith a quick text: "What are you doing tonight? Want to meet up? Got a costume?"

The prosecutor and the press later used Meredith's reply, "Yes I have one, but I have to go to a friend's house for dinner. What are your plans?" as proof of our fraying relationship, even though she signed off with an *X* for kisses. But Meredith had her own set of friends, and I didn't expect to be included in everything she did. I texted my friend Spyros, the guy who worked at the Internet café, and we agreed to catch up.

Le Chic, usually so empty and desolate feeling, was packed that evening, possibly for the first time in its yearlong existence. Juve was standing near the front door. He'd painted his face white and had fake blood trickling down from the corner of his mouth. "Where's your costume?" he asked.

"I'm a kitty cat."

"You're supposed to be scary," Juve said.

Patrick poured me a glass of wine, and I hung out on the edge of the crowd for a while. But, for some reason, I was feeling a bit flat. I caught Patrick's eye and mouthed, "I'm leaving," waving good-bye. He gave me a nod, and I was out the door.

Around 12:30 A.M., when I met Spyros and his friends for drinks, I couldn't get into the good time they were having. Even on a blowout party night, Perugia's social scene didn't do much for me, and the whole evening felt like a dud. It made me nostalgic for the sit-around-and-talk gatherings of friends at UW. I was glad when Raffaele came to Piazza IV Novembre to walk me home. By that time it was 1:45 A.M., and most of my eyeliner whiskers had rubbed off. Thankfully, Halloween 2007 was over.

A ll Saints' Day, on November 1, is a national holiday in Italy, a day to honor the dead. Except for church bells ringing, Perugia was quiet that Thursday morning. Everyone my age must have been sleeping off the night before. I was happy for the solitude when I left Raffaele's for a few hours on my own at home.

Around noon, I was sitting at the kitchen table reading when Filomena and her boyfriend, Marco, stopped by to change clothes for a party. She was in a rush as she chatted with me through her open bedroom door.

"How are you?" she asked. "Where's Meredith?"

"I'm good," I said. "Just waiting for Raffaele to come over for lunch. Meredith must still be asleep."

Filomena and Marco left about an hour before Meredith wandered in from her room. She looked sleepy-eyed.

"You've still got vampire blood on your chin," I told her.

"I know. I couldn't get all the paint off," she said. "I was so tired when I got home at five A.M. that I didn't wash my face."

"What did you end up doing last night?" I asked.

"I went to a dinner party. It was amazing. They filled a surgical glove with water and froze it to make an ice hand. It looked cool floating in the punch bowl. Then we all went dancing at Merlin's"—Meredith's favorite pub. "What about you?"

"My Halloween was lame. I thought it was fun seeing everyone's costumes, but mostly I was bored."

By the time Meredith got out of the shower, Raffaele was at our house. We were eating pasta when she came out of her room carrying an armload of dirty clothes to put in the washer in the big bathroom. She was wearing baggy boyfriend-style jeans.

"Dang, girl. Nice pants," I said.

"Yeah, my ex-boyfriend bought them for me," she said, hip-bumping me in response to my compliment. "What are you doing for the rest of the day?"

"Hanging out here for a while and then we're going back to Raffaele's," I answered.

"Oh, cool. I'm heading out with friends, so have a good day."

She grabbed her purse. "See you. *Ciao*," she said, tossing the strap over her shoulder and waving as she went through the front door.

Raffaele and I were good at being low-key together. We chilled out in the common room and smoked a joint while I played Beatles songs on the guitar for an hour or so. Sometime between 4 P.M. and 5 P.M., we left to go to his place. We wanted a quiet, cozy night in. As we walked along, I was telling Raffaele that *Amélie* was my all-time favorite movie.

"Really?" he asked. "I've never seen it."

"Oh my God," I said, unbelieving. "You have to see it right this second! You'll love it!"

Not long after we got back to Raffaele's, his doorbell rang. It was a friend of his whom I'd never met—a pretty, put-together medical student named Jovanna Popovic, who spoke Italian so quickly I couldn't understand her. She'd come to ask Raffaele for a favor. Her mother was putting a suitcase on a bus for her and she wondered if he could drive her to the station at midnight to pick it up.

"Sure," Raffaele said.

As soon as she left, we downloaded the movie on his computer and sat on his bed to watch it. Around 8:30 P.M. I suddenly remembered that it was Thursday, one of my regular workdays.

Quickly checking my phone, I saw that Patrick had sent me a text telling me I didn't have to come in. Since it was a holiday, he thought it would be a slow night.

"Okay," I texted back. *"Ci vediamo più tardi buona serata!"*— "See you later. Have a good evening!" Then I turned off my phone, just in case he changed his mind and wanted me to come in after all. I was so excited to have the night off that I jumped on top of Raffaele, cheering, "Woo-hoo! Woo-hoo!"

Our good mood was only elevated when the doorbell rang again at 8:45 P.M.: Jovanna had come back, this time to say that the suitcase hadn't made the bus and that she didn't need a ride after all. With no more obligations, we had the whole rest of the night just to be with each other and chill out.

After the movie ended, around 9:15 P.M., we sautéed a piece of fish and made a simple salad. We were washing the dishes when we realized that the kitchen sink was leaking. Raffaele, who'd already had a plumber come once, was frustrated and frantically tried to mop up a lot of water with a little rag. He ended up leaving a puddle.

"I'll bring the mop over from our house tomorrow. No big deal," I said.

Raffaele sat down at his desk and rolled a joint, and I climbed into his lap to read aloud to him from another Harry Potter book, this one in German. I translated the parts he didn't understand, as best I could, into Italian or English while we smoked and giggled.

Later, when we were in bed, our conversation wound its way to his mother. His dad had divorced her years before, but she'd never gotten over the break. In 2005 she had died suddenly. "Some people suspect she killed herself, but I'm positive she didn't," Raffaele said. "She would never do that. She had a bad heart, and it just gave out. It was horrible for me—we were really close—and I miss her all the time."

I felt terrible for him, but it was hard for me to relate. The only person I knew who had died was my grandfather, when I was sixteen. I felt sad when my mom told me, but my grandfather had been old and sick, and we had expected his death for a few weeks.

I'm sure Mom and Oma must have cried, but my strongest memory is sitting around the dining room table telling funny stories about Opa. My grandmother's message—that grieving was something you did in private; that you didn't make public displays and you kept on moving forward—had remained with me.

Hearing the pain in Raffaele's voice, I hurt for him. Nestling my head on his chest, I tried to be comforting.

As we started kissing, Raffaele gave me a hickey on my neck. We undressed the rest of the way, had sex, and fell asleep.

We'd known each other for exactly one week and had settled so quickly into an easy routine that one night seemed to melt happily and indistinguishably into the one that came after.

We planned to break our routine the next day, All Souls' Day, by taking a long drive into the countryside, to the neighboring town of Gubbio. The November 2 holiday wasn't usually observed with as much fanfare as All Saints' Day, but since it fell on a Friday in 2007, a lot of people, including us, were turning it into a four-day weekend. I thought, *Italians having a good time again.* And I couldn't wait.

Chapter 6

Morning, November 2, 2007, Day One

On that cold, sunny Friday morning, I left Raffaele asleep in his apartment and walked home to take a shower and get my things together, thinking about our romantic weekend in the Umbrian hills. In hindsight, it seems that arriving home to find the front door open should have rattled me more. I thought, *That's strange.* But it was easily explained. The old latch didn't catch unless we used a key. *Wind must have blown it open,* I thought, and walked inside the house calling out, "Filomena? Laura? Meredith? Hello? Hello? Anybody?"

Nobody. The bedroom doors were closed.

I wasn't alarmed by two pea-size flecks of blood in the bathroom sink that Meredith and I shared. There was another smear on the faucet. *Weird.* I'd gotten my ears pierced. Were they bleeding? I scratched the droplets with my fingernail. They were dry. *Meredith must have nicked herself.*

It wasn't until I got out of the shower that I noticed a reddish-brown splotch about the size of an orange on the bathmat. *More blood. Could Meredith have started her period and dripped? But then, how would it have gotten on the sink?* My confusion increased. We were usually so neat. I went to my room and, while putting on a white skirt and a blue sweater, thought about what to bring along on my trip to Gubbio with Raffaele.

I went to the big bathroom to use Filomena's blow dryer and was stashing it back against the wall when I noticed poop in the toilet. No one in the house would have left the toilet unflushed. *Could there have been a stranger here? Was someone in the house when I was in the shower?* I felt a lurch of panic and the prickly feeling you get when you think someone might be watching you. I quickly grabbed my purse and coat and somehow remembered the mop I said I'd bring back to Raffaele's. I scrambled to push the key into the lock, making myself turn it before I ran up the driveway, my heart banging painfully.

By the time I was a block from home I was second-guessing myself. Maybe I was overreacting. Maybe there was a simple reason for the toilet being unflushed. I needed someone to say, "Amanda, you're right to be scared. This isn't normal." And if it wasn't okay, I wanted someone to tell me what to do. My skittering brain pulled up my mom's mantra: when in doubt, call. Forgetting the nine-hour time difference between Perugia and Seattle, I pressed the number sequence for home. My mom did not say hello, just "Amanda, are you okay? What's wrong?" It was in the middle of the night in Seattle, and she was worried.

"I'm on my way back to Raffaele's," I said, "but I just wanted to check in. I found some strange things in my house." I explained my reasons for worrying. Then I asked, "What do you think I should do?"

"Call your roommates," she said. "Go tell Raffaele, and call me right back."

Hearing Mom's voice calmed me. *It can't be that bad*, I thought. *I'm out of the house. Nothing happened. I'm safe. No one's in danger.*

I called Filomena first and was relieved when she picked up.

"*Ciao*, Amanda," she said.

"*Ciao*," I said. "I'm calling because when I came home from Raffaele's this morning, our front door was open. I found a few drops of blood in one bathroom and shit in the other toilet. Do you know anything about it?"

"What do you mean?" she asked, her voice instantaneously on high alert. "I didn't stay there last night—I was at Marco's—and Laura's in Rome on business. Have you talked to Meredith?"

"No, I tried you first," I said.

"I'm at the fair outside town," she said. "I just got here. Try Meredith, and then go back to the house. We need to see if anything was stolen." She sounded worried.

I called Meredith on her British phone. A recording said it was out of service. That struck me as odd. Then I pulled up Meredith's Italian number. It went straight to voice mail.

By that time, I was back at Raffaele's. He was in total vacation mode: he'd slept in and had just gotten out of the shower. I'd forgotten about our trip. "Hey," I said, trying to sound casual, "does this sound weird to you?" I told him what I'd seen.

"Yeah," he said. "We should definitely go over and look around."

Over a quick breakfast, Raffaele and I talked some more about what I'd seen. "Maybe the toilet is just broken," he said.

Even before we'd downed the last sips of our coffee, Filomena called back. "What do you see?" she demanded. Her panic was retriggering my own.

"Filomena," I said, as evenly as I could, "we're just leaving Raffaele's."

Amanda Knox

Ten minutes later, when we reached the villa, my stomach was knotted with dread. "What if someone was in here?" I said, feeling increasingly creeped out. Raffaele held my free hand while I unlocked the door. I yelled, "Is anyone here?"

At first nothing seemed amiss. The house was quiet, and the kitchen/living area was immaculate. I poked my head in Laura's room. It looked fine, too. Then I opened Filomena's door. I gasped. The window had been shattered and glass was everywhere. Clothes were heaped all over the bed and floor. The drawers and cabinets were open. All I could see was chaos. "Oh my God, someone broke in!" I shouted to Raffaele, who was right behind me. In the next instant, I spotted Filomena's laptop and digital camera sitting on the desk. I couldn't get my head around it. "That's so weird," I said. "Her things are here. I don't understand. What could have happened?"

Just then, my phone rang. It was Filomena. "Someone's been in your room," I said. "They smashed your window. But it's bizarre—it doesn't look like they took anything."

"I'm coming home this second," she said, her voice constricted.

Meredith's door was still closed, just as it had been when I was home earlier. I called out, "Meredith." She didn't answer. *Could she have spent the night with Giacomo? Or with one of her British girlfriends?* Still, at that moment I was more worried about the smashed window in Filomena's room than about Meredith's closed door.

I ran outside and around the house to see if the guys downstairs were home and to see if they'd heard anything during the night. Outside, away from Raffaele, my anxiety soared. My heart started racing again. I pounded on their door and tried to peer through the glass. It looked like no one was home.

I ran back upstairs and knocked gently on Meredith's door, calling, "Meredith. Are you in there?" No sound. I called again, louder. I knocked harder. Then I banged. I jiggled the handle. It was locked. *Meredith only locks her door when she's changing clothes*, I thought. *She can't be in there or she'd answer.* "Why isn't she answering me?" I asked Raffaele frantically.

I couldn't figure out, especially in that moment, why her door would be locked. What if she were inside? Why wouldn't she respond if she were? Was she sleeping with her earphones in? Was she hurt? At that moment what mattered more than anything was reaching her just to know where she was, to know that she was okay.

I kneeled on the floor and squinted, trying to peer through the keyhole. I couldn't see anything. And we had no way of knowing if the door had been locked from the inside or the outside.

"I'm going outside to see if I can look through her window from the terrace."

I climbed over the wrought-iron railing. With my feet on the narrow ledge, I held on to the rail with one hand and leaned out as far as I could, my body at a forty-five-degree angle over the gravel walkway below. Raffaele came out and shouted, "Amanda! Get down. You could fall!"

That possibility hadn't occurred to me.

"Please come in before you get hurt!"

As soon as we got inside, we went back to Meredith's closed door. "I can try to kick it down," Raffaele offered.

"Try it!"

He rammed the door with his shoulder, hard. Nothing. He kicked next to the handle. It didn't budge.

I called my mom again. "Mom," I said. "Someone broke into our house, and we can't find Meredith. What should we do?"

"Amanda, call the police," she said.

My stepfather, Chris, yelled into the speakerphone, "Amanda, get the hell out of the house, this instant!"

While I was talking to them, Raffaele called his sister to see what she thought. She was a police officer in Rome.

Raffaele dialed 112—Italy's 911—for the Carabinieri, which was separate from—and more professional than—the Perugian town police.

As soon as he hung up, I said, "Let's wait for them outside." Even without Chris's insistence, I was too spooked to be in the house. On the way out I glanced from the kitchen into the larger bathroom. The toilet had been flushed. "Oh my God!" I said to Raffaele. "Someone must have been hiding inside when I was here the first time—or they came back while I was gone!"

We ran out and waited on a grassy bank beside the driveway. I was shivering from nerves and cold, and Raffaele was hugging me to calm me down and keep me warm, when a man in jeans and a brown jacket walked up. As he approached us he said he was from the police. I thought, *That was fast.*

Another officer joined him. I tried to explain in Italian that there had been a break-in and that we hadn't been able to find one of our roommates, Meredith. With Raffaele translating both sides, I gradually understood that these officers were just Postal Police, the squad that deals with tech crimes.

"Two cell phones were turned in to us this morning," one said. "One is registered to Filomena Romanelli. Do you know her?"

"Yes, she's my housemate," I said. "It can't be Filomena's, because I just talked to her. But I've been trying to reach my other roommate, Meredith, all morning. She doesn't answer. Who turned these in? Where did they find them?"

Later I found out that a neighbor had heard the phones ringing in her garden when I'd tried to call Meredith. They'd been tossed over the high wall that protected the neighbor's house from the street—and from intruders. But the Postal Police wouldn't explain or answer my questions.

We went inside, and I wrote out Meredith's phone numbers on a Post-it Note for them. While we were talking, we heard a car drive up. It was Filomena's boyfriend, Marco Z., and his friend Luca. Two minutes later, another car screeched into the driveway—it was Filomena and her friend Paola, Luca's girlfriend. They jumped out, and Filomena stormed into the house to scavenge through her room. When she came out, she said, "My room is a disaster. There's glass everywhere and a rock underneath the desk, but it seems like everything is there."

The Postal Police showed her the cell phones. "This one is Meredith's British phone," Filomena said. "She uses it to call her mother. And I lent her the SIM card to the other one to make local calls."

The men seemed satisfied; their work was done. They said, "We can make a report that there's been a break-in. Are you sure nothing was stolen?"

"Not as far as we can tell," I said. "But Meredith's door is locked. I'm really worried."

"Well, is that unusual?" they asked.

I tried to explain that she locked it sometimes, when she was changing clothes or was leaving town for the weekend, but Filomena wheeled around and shouted, "She *never* locks her door!" I stepped back and let her take over the conversation, Italian to Italian. The rapid-fire exchange stretched way past my skills. Filomena shouted at the Postal Police officers, "Break down the door!"

"We can't do that; it's not in our authority," one said.

Six people were now crammed into the tiny hallway outside Meredith's bedroom, all talking at once in loud Italian. Then I heard Luca's foot deliver a thundering blow. He kicked the door once, twice, a third time. Finally the impact dislodged the lock, and the door flew open. Filomena screamed, *"Un piede! Un piede!"*—"A foot! A foot!"

A foot? I thought. I craned my neck, but because there were so many people crowding around the door, I couldn't see into Meredith's room at all. "Raffaele," I said. He was standing beside me. "What's going on? What's going on?"

One of the guys shouted, *"Sangue! Dio mio!"*—"Blood! My God!"

Filomena was crying, hysterical. Her screams sounded wild, animal-like.

The police boomed, "Everyone out of the house. Now!" They called for reinforcements from the Perugian town police.

Raffaele grabbed my hands and pulled me toward the front door.

Sitting outside on the front stoop, I heard someone exclaim, *"Armadio"*—"armoire." *They found a foot in the closet,* I thought. Then, *"Corpo!"*—"A body!" *A body inside the wardrobe with a foot sticking out?* I couldn't make the words make sense. Filomena was wailing, "Meredith! Meredith! Oh, God!" Over and over, "Meredith! Oh, God!"

My mind worked in slow motion. I could not scream or speak. I just kept saying in my head, *What's happening? What's happening?*

It was only over the course of the next several days that I was able to piece together what Filomena and the others in the doorway had seen: a naked, blue-tinged foot poking out from beneath Meredith's comforter, blood splattered over the walls and streaked across the floor.

But at that moment, sitting outside my villa, the image I had was of a faceless body stuffed in the armoire, a foot sticking out.

Maybe that's why Filomena cried, and I didn't. In that instant, she'd seen enough to grasp the terrible scope of what had happened. All I got was confusion and words and, later, question after question about Meredith and her life in Perugia. There was nothing I could say about what her body was like in its devastation.

But even with all these blanks, I was still shaken—in shock, I'd guess. Waiting in the driveway, while two policemen guarded the front door, I clung to Raffaele. My legs wobbled. The weather was sunny, but it was still a cold November day, and suddenly I was freezing. Since I'd left the house without my jacket, Raffaele took off his gray one with faux-fur lining and put it on me.

Paramedics, investigators, and white-suited forensic scientists arrived in waves. The police wouldn't tell us anything, but Luca and Paola stayed close, trying to read lips and overhear. At one point, Luca told Raffaele what the police had said: "The victim's throat has been slashed."

I didn't find out until the months leading up to my trial—and during the trial itself—how sadistic her killer had been. When the police lifted up the corner of Meredith's beige duvet they found her lying on the floor, stripped naked from the waist down. Her arms and neck were bruised. She had struggled to remain alive. Her bra had been sliced off and left next to her body. Her cotton T-shirt, yanked up to expose her breasts, was saturated with blood. The worst report was that Meredith, stabbed multiple times in the neck, had choked to death on her own blood and was found lying in a pool of it, her head turned toward the window, eyes open.

In the first hours after the police came, standing outside the villa that had been the happy center of my life in Perugia—my refuge thousands of miles from home—I mercifully didn't know any of this. I was slowly absorbing and rejecting the fractured news that Meredith was dead.

I felt as if I were underwater. Each movement—my own and everyone else's—seemed thick, slow, surreal. I willed the police to be wrong. I wanted Meredith to walk down the driveway, to be alive. What if she'd spent the night with one of her British girlfriends? Or gotten up early to meet friends? I held the near-impossible idea that somehow the person in Meredith's room was a stranger.

Nothing felt real except Raffaele's arms, holding me, keeping me from collapsing. I clung to him. Unable to understand most of what was being said, I felt cast adrift. My grasp of Italian lessened under the extraordinary stress. Catching words and translating in my head felt like clawing through insulation.

I was flattened. I was in despair. I cried weakly on and off into Raffaele's sweater. I never sobbed openly. I'd never cried publicly. Perhaps like my mom and my Oma, who had taught me to cry when I was alone, I bottled up my feelings. It was an unfortunate trait in a country where emotion is not just commonplace but expected.

Raffaele's voice was calm and reassuring. *"Andrà tutto bene"*— "It's going to be okay," he said. He pulled me closer, stroked my hair, patted my arm. He looked at me and kissed me, and I kissed him back. These kisses were consoling. Raffaele let me know that I wasn't alone. It reminded me of when I was young and had nightmares. My mom would hold me and smooth my hair and let me know that I was safe. Somehow Raffaele managed to do the same thing.

Later, people would say that our kisses were flirtatious— evidence of our guilt. They described the times I pressed my face to Raffaele's chest as snuggling. Innocent people, the prosecutor and media said, would have been so devastated they'd have been unable to stop weeping.

Watching a clip of it now, my stomach seizes. I'm gripped by the same awful feelings I had that afternoon. I can only see myself as I was: young and scared, in need of comfort. I see Raffaele trying to cope with his own feelings while trying to help me.

We waited in the driveway for what seemed like forever. The police officers would come out, ask us questions, go in, come out, and ask more questions. I always told them the same thing: "I came home. I found the door open. Filomena's room was ransacked, but nothing seems to have been stolen. Meredith's door was locked."

It seemed like the words came from somewhere else, not from my throat.

In the middle of my muddy thoughts I had one that was simple and clear: "We have to tell the police that the poop was in Filomena and Laura's bathroom when I put the hair dryer away and was gone when we came back," I told Raffaele. The poop must have belonged to the killer. *Was he there when I took my shower? Would he have killed me, too?*

We walked up to a female officer with long black hair and long nails—Monica Napoleoni, head of homicide, I later found out. Raffaele described in Italian what I'd seen. She glared at me. "You know we're going to check this out, right?" she said.

I said, "That's why I'm telling you."

She disappeared into the villa, only to return moments later. "The feces is still there. What are you talking about?" she spat.

This confused me, but I continued to tell her what happened anyway. I told her I'd taken the mop with me in the morning but had brought it back when Raffaele and I came to see if the house had been robbed.

"You know we're going to check that for blood, too?" she asked.

"Okay," I said. I was surprised by how abrupt she was.

The police explained that they couldn't let us back into the house, that it would compromise the crime scene. Before we were told to go outside, Filomena had carefully gone through her room to see if anything had been stolen. Now, having calmed down momentarily, she came over and whispered that she couldn't leave without her laptop, that she had to have it for work. She snuck back into her room—I have no idea how she got past the police standing sentry—and grabbed it, disturbing the scene for a second time. Marco stood in the driveway, looking lost. Paola and Luca had slipped off to the car, where it was warm.

Word was already spreading. I'm not sure how. A few people texted me, asking, "What's going on?" I texted back that, yes, it was our villa. Yes, unbelievably, it was Meredith. I noticed that TV crews had set up in the parking lot above our house. But their presence in the distance barely registered with me.

Sometime around 3 P.M. the police gathered us all in the driveway and told us to meet them at the station—the *questura*—a tall, generic, off-white modern building about ten minutes away, on the edge of town.

Raffaele and I rode with Luca and Paola. During the drive I was again overcome by the realization that the body had to be Meredith's, that she was dead. I couldn't push it away. I doubled over in the backseat and sobbed. Raffaele put his hand on my back, and Paola looked over and said, "It's okay, it's okay."

But it wasn't okay.

_____

Chapter 7

Afternoon, November 2, 2007, Day One

At the police station, one question hung in the air: Who? Who could have done this?

As the police started trying to figure out the answer, it made sense that they started their questioning with me. I was the first person to come home that morning. I was anxious to explain everything I'd noticed, starting with the open front door and the droplets of blood in the sink.

They led me through the waiting room into a generic office—long and narrow, with a small window at the far end. For the first hour, I was questioned in Italian, but it was so hard for me to follow and explain that they brought in an English-speaking detective for hours two through six. Alone in the room, we sat on opposite sides of a plain wooden desk. I described everything I could think of. Some questions he asked were obvious. Others seemed irrelevant. "*Anything* might be a clue for the investigators," he said. "Don't hold back—even if it seems trivial. The smallest detail is important. You never know what the key will be to finding the person who did this."

He asked, and I rushed to answer.

"How did you meet Meredith? How long have you been in Perugia? Who was Meredith dating? What do you know about the guys who live downstairs? Where did Meredith like to party? When was the last time you saw her? Where was she going? What time did Meredith leave home?"

"It was yesterday afternoon. I don't know where she was heading," I said. "She didn't tell us."

"What did you and Raffaele do yesterday afternoon and last night?" he asked.

"We hung out at my house and then at Raffaele's apartment."

He didn't press me. He just listened.

It seemed like a straightforward debriefing. I was too naïve to imagine that the detectives suspected that the murder had been an inside job and that the burglary had been faked. I had no way of knowing that the Postal Police had thought Raffaele's and my behavior suspicious. The detective didn't say any of this. Nor did he allow that the homicide police had begun to watch us closely before we'd even driven out of the driveway.

Now I see that I was a mouse in a cat's game. While I was trying to dredge up any small thing that could help them find Meredith's killer and trying to get my head around the shock of her death, the police were deciding to bug Raffaele's and my cell phones.

As I sat waiting to hear what else the police needed from me, I asked the detective if it was true that it was Meredith who had been murdered. I still couldn't let go of the tiniest hope that the body in her room hadn't been Meredith's, that she was still alive.

The detective nodded and ran his finger in a cutting motion across his neck.

I covered my mouth with my hands and shook my head back and forth. *No.* "I just can't believe it," I said softly.

He nodded again, soberly, looking me in the eye.

After that I was sent to the waiting room. It was crowded with Meredith's friends. Everyone was crying, talking, milling around the police station, trying to make sense of the senseless. Meredith's British girlfriends, including Sophie, Amy, and Robyn, with whom I'd eaten dinner my first night in town, sat together. The owner of Merlin's, Meredith's favorite bar, was there. Laura had been called back from Rome. The guys from the downstairs apartment had been spending the holiday weekend a few hours away, in their hometown. They were taking the train back to Perugia when they got a call about Meredith and were told to come to the *questura*. Giacomo looked stricken.

Everyone had questions: "What did you see? What do you know?"

Trying to be helpful, I shared the information I had, much of which turned out to be wrong. I still thought Meredith's body had been found stuffed into the armoire.

When I first saw Laura, she was dry-eyed. She came up and hugged me and said, "I can't believe it. I'm so sorry. I know Meredith was your friend." Then she sat me down and said, "Amanda, this is really serious. You need to remember: do not say anything to the police about us smoking marijuana in our house."

I was thinking, *You can't lie to the police*, but I considered this anxiously a moment and then said, "Okay, I haven't yet. I won't."

I asked, "Do you think they'll let us get our stuff out of the house?"

Laura said, "I hope so. Filomena and I are talking to our lawyers about that."

It didn't occur to me—or to my parents, who were now calling me nonstop—that perhaps I should call a lawyer, too.

Meredith's British friends were huddled together. Sophie came up and gave me a one-sided hug. I was too wrung out at that moment to reciprocate. This was the first of many things I did, in just being myself, that didn't play well for me in this stressed environment. Along with my lack of tears, this uncommon behavior was used to show that I had no feelings for Meredith. Sophie then explained to me that she had walked Meredith most of the way home the previous night and so was the last person to see her alive.

Around 3 A.M. a police officer led the British girls and me downstairs to get fingerprinted. "We need to know which fingerprints to exclude when we go through the house," he said.

One by one they took us into a room and painted our fingertips with a black, tarlike syrup. When I came out, Sophie was sitting on a chair outside the door, sobbing. I tried to make up for my earlier lack of warmth, saying, "I'm so sorry about Meredith. If you need anything, here's my number."

And suddenly, I woke up from deep shock. I was struck with righteous fury against Meredith's murderer. I started pacing the hallway. I was so outraged I was shaking and hitting my forehead with the heel of my palm, saying, "No, no, no," over and over. It's something I've always done when I can't contain my anger.

The English-speaking detective who'd been overseeing the fingerprinting approached me and said, "Amanda, you need to calm down."

I felt an overpowering need to help track down the murderer. I wanted to make sure he—I assumed it was a he—spent the rest of his life in prison. I wanted him to regret his wrongdoing every hour of every day. Forever.

I kept thinking, *How could this happen? How could it have been Meredith? She wasn't a person who made enemies.* She'd been murdered in our house, in her own bedroom. She was at home, where she should have been safe. I felt sick. The body's primal response to stress is to fight or flee, and at that moment I was all fight.

What if my natural response had been to leave? In the days that followed it didn't occur to me that I actually could have returned home to recuperate and contribute to the investigation from afar, or that it could have been the right thing to do.

As I continued walking back and forth in the hallway, my mind kept looping back around itself, making quick, tight turns: *What happened? Who would leave poop in the toilet? Why hadn't Laura's and my rooms been touched? Why was Filomena's computer still there? Did Meredith know her attacker? How could this have happened? How? How? How?*

I tried to sit back down next to Sophie, but I couldn't stay still. I said, "I can't imagine who would do this. It just does not make any sense to me."

The detective, who was still standing there, said, "We're going to try to find out as fast as we can, and anything you can remember will help us."

When we went back upstairs to the waiting room my family called me, one after another. I have a tendency to talk too loudly when I'm excited. That night, explaining to them what had happened, wasn't an exception. But I didn't notice this. I had to repeat my story for each caller—how I'd been the first one home and hadn't realized there had been a break-in, how Raffaele and I had called the police.

My stepdad, Chris, said, "The killer might have been watching your house for days and saw that Meredith was alone. He might know where Raffaele lives. You need to be careful! Pay attention to what's going on around you. Make sure you're with someone all the time."

My dad said, "I wish I could put my arms around you and protect you." He asked to speak to Raffaele.

My dad doesn't speak two words of Italian, but Raffaele understood and agreed when Dad said, "Raffaele, thank you for taking care of Amanda for me. Please make sure to keep her safe."

When I wasn't on the phone, I paced. I walked by one of Meredith's British friends, Natalie Hayworth, who was saying, "I hope Meredith didn't suffer."

Still worked up, I turned around and gaped. "How could she not have suffered?" I said. "She got her fucking throat slit. Fucking bastards."

I was angry and blunt. I couldn't understand how the others remained so calm. No one else was pacing. No one else was muttering or swearing. Everyone else was so self-contained. First I showed not enough emotion; then I showed too much. It's as if any goodwill others had toward me was seeping out like a slow leak from a tire, without my even realizing it.

I suspect that Raffaele thought I was having a breakdown. He sat me in his lap and bounced me gently. He kissed me, made faces at me, and told me jokes—all in an effort to soothe my agitation, babying me so I would stop storming around.

I cringe to say that treating me like an infant helped. Normally it would have repelled me. But at that time it worked.

Finally I took my journal from my purse and scribbled down a few stream-of-consciousness lines about how unreal all of this was and how I wished I could write a song about the heinous, tragic event—a personal tribute to Meredith. I thought that, like the act of writing itself, music might somehow help me feel better. Later, when the police confiscated my notebook and its contents were leaked to the press, people saw this as proof that I was trivializing Meredith's death.

They found more evidence in my gallows humor. I wrote, "I'm starving. And I'd really like to say that I could kill for a pizza but it just doesn't seem right."

I had so many thoughts clamoring in my brain at once that I was writing whatever came into my head. I never meant to share these things, only to give myself some relief. The words in my journal were taken literally, and they damned me. It was a situation I would find myself in again and again.

It was early morning by the time I put my notebook away. The police weren't stopping to sleep and didn't seem to be allowing us to, either. Raffaele and I were part of the last group to leave the *questura*, along with Laura, Filomena, Giacomo, and the other guys from downstairs, at 5:30 A.M.

The police gave Raffaele and me explicit instructions to be back at the *questura* a few hours later, at 11 A.M. "Sharp," they said.

I can't recall who dropped us off at Raffaele's apartment. But I do remember being acutely aware that I didn't have anywhere else to go.

Chapter 8

November 3, 2007, Day Two

W hen Raffaele and I returned to the *questura* on Saturday, at 11 A.M., the waiting room was empty. Some of Meredith's British friends were flying home that day, too devastated and scared to stay. Two of the girls had caught a bus to the airport at seven o'clock—about the time I was finally just getting to sleep.

I had the same opportunity. Mom had asked in one of our phone conversations the night before if I wanted her to buy me a plane ticket to Seattle. "No," I said. I had been adamant. "I'm helping the police."

I never considered going home. I didn't think it was right to run away, and that's exactly how I looked at it—as running away from being an adult. I knew that murders can and do happen anywhere, and I was determined not to let this tragedy undo all I'd worked so hard for over the past year. I liked my classes at the University for Foreigners, and I knew my family's finances didn't allow for re-dos. The way I saw it, if I went home, I'd be admitting defeat. And my leaving wouldn't bring Meredith back.

But I understood why Meredith's British girlfriends were

panic-stricken. I was, too. That morning a London newspaper had called Meredith's killer "a knife-wielding maniac." He was still on the loose, possibly getting ready to strike other victims, possibly me. I didn't need Chris to warn me against being alone. I was already so paranoid I refused to let Raffaele out of sight in his one-room apartment. Walking down the street with his arm around me, I kept looking nervously over my shoulder to make sure no one was following us. Passing cars made me jump. Had the murderer watched our house, waiting until one of us was alone to make his move? I couldn't help but wonder, *Would I have died if I'd been home Thursday night?* All that separated Meredith's and my room was one thin wallboard. *Why am I alive and she's now lying in the morgue?* And: *Could I be the next victim?*

I hated that I felt so traumatized. As my family, friends, and the UW foreign exchange office checked in one after another, they each said some version of "Oh my God, you must be so scared and alone." I didn't want to admit that they were right, that what I was going through was too stressful for me to handle by myself. But the last thing I wanted from my parents—even though it's probably what I needed most—was to be treated like a child.

I believed I had to demonstrate to Mom, Dad, and myself—as if my whole personhood depended on it—that I was in control, that I could take care of things in a mature, responsible way. And just as I'd had some wrong-headed notion about the link between casual sex and adulthood, I was also sure that an adult would know how to deal with whatever was thrown at her— including how to behave if her roommate were brutally murdered. It wasn't logical, but I believed that I'd made the decision to come to Perugia and that, while no one could possibly have anticipated Meredith's death, I just had to suck it up. I treated the whole incident as if it were an unanticipated situation I had found myself in and now I had to handle it.

So, anytime I was on the phone with my parents I put my energy into reassuring them that I was okay. Just as I hadn't wanted to alarm my mom when I'd first run out of the villa after seeing the poop in the toilet, I still didn't want to alarm her. Therefore, each phone conversation was more or less the same. "Yeah, I'm really tired, but it's going to be okay. I'm with Raffaele. He's taking good care of me. My roommates are looking for a new place. Don't worry, don't worry, don't worry."

We don't have a traditional parent-child relationship, one where they would have insisted I come home against my wishes. And, at the time, I believed what I was telling them. But looking back now, I think I was too afraid to admit the truth, that it would somehow mean I'd failed.

On a Saturday morning when I would ordinarily have been drinking coffee in my pajamas and reading my Italian *Harry Potter*, I was sitting in a sterile police office waiting to be questioned—again. I was wearing the same clothes I'd put on for my date with Raffaele a day earlier; they were now all I had. I'd hardly slept in twenty-four hours. Nor had I been able to quell the mind-flattening rage that had erupted the previous night. The only way to do that, I was sure, was to help the police find Meredith's killer. I wanted justice for her, and as her friend, roommate, and the person who might also have been murdered had I been there that night, I was sure I was the police's best resource.

The police immediately sent Raffaele home and sat me down in front of an old computer monitor to identify who was who in

Facebook pictures of Meredith and her friends on Halloween. I didn't know many of her buddies to begin with, but the job of figuring out people's identities had been rendered nearly impossible by the fact that almost everyone, including Meredith, was wearing a lurid disguise—zombie paint, *Scream* masks, fake teeth, vampire blood. The irony was painful.

When we finished, a detective put me through a second round of questioning, this time in Italian. Did we ever smoke marijuana at No. 7, Via della Pergola? "No, we don't smoke," I lied, squirming inwardly as I did.

I didn't see that Laura had left me with any choice, and I felt completely trapped by her demand. I could barely breathe until the detective moved on to a new topic, and when he did, I was hugely relieved. I thought that was the end of it.

Even through the language barrier, I picked up on a change in the detective's tone from the night before. He was pushy, his questions repetitive. He told me to list the people who'd visited our house, and any guys Meredith knew. "We need every name," he said. "Who invited them? How many times did they come over? What type of relationship did Meredith have with them? Did she ever have a fight with them?"

Aside from what I said about our villa's drug habits, I told him everything I could possibly think of. I scoured my brain to remember anyone who had even glanced at Meredith. I scrolled through my Italian phone and gave him the names and numbers of every contact I had. Even with all that, he acted as if what I'd told him wasn't enough. He kept pressing for more. I didn't have any more.

It's hard to believe I had no inkling that the police suspected me. But why would they have? I was innocent. I'd been taught by my parents to do my civic duty. I was so intent on helping them, I couldn't step back. And I thought I understood why they were pressuring me.

If you drew a diagram with Meredith's housemates in one circle and her friends in another, I was the only person in Italy in both circles. Unlike Laura and Filomena, Meredith and I were close in age, both college students, and native English speakers. Unlike Amy, Robyn, and the other British girls, we spit out our toothpaste into the same sink and shared the food in the fridge. If anyone knew a detail that could help track down her killer, it would likely have been me.

When I wasn't being questioned, I hung out in the waiting room for the police to give me a new set of instructions. I spent almost every free minute on the phone with one of my parents. Mom and Aunt Dolly had decided it would be good for me to spend some time with Dolly and her family in Hamburg until the murderer was caught. I was willing to go anywhere, as long as it wasn't home for good.

That afternoon, I was talking to a steely brown-haired police officer named Rita Ficarra—although I didn't find out her name until two years later, when she testified against me in court. I said, "My parents want me to go to Germany to stay with relatives for a couple of weeks. Is that okay?"

She said, "You can't leave Perugia. You're an important part of the investigation."

She didn't seem like a person you'd ever want to argue with. "How long will you need me?" I asked.

"We don't know—maybe months," she said.

This stunned me. "But I'm planning to go home for Christmas."

"Well, we'll decide if you can do that," she said. "We'll have to hear what the magistrate says when he calls in three days."

When I repeated this conversation to my mom, she was concerned. "That doesn't make sense," she said.

Later that afternoon, Mom asked, "Amanda, do you need me there?"

Although it had been only one day since Meredith's body had been discovered, I said, "I know I'll be okay, but I'd really appreciate it if you came."

My chest loosened when she called back again with her flight information. She was due to land in Rome on Tuesday morning, November 6. From there she'd take a train to Perugia, and I'd meet her at the station. She said she wanted to help me find a new place to live and buy me some clothes. I looked at her visit as a way to get my life back on track.

Sometime that afternoon the police drove me to the villa. Sitting in the backseat with an interpreter on the way there, I admitted, "I'm completely exhausted."

One of the officers in the front seat swung around and looked at me. Her reaction was harsh: "Do you think we're not tired? We're working twenty-four/seven to solve this crime, and you need to stop complaining. Do you just not care that someone murdered your friend?"

The police told me to cover myself in the car as we neared the villa, so that the satellite trucks and photographers who'd commandeered the parking lot above us, their cameras aimed right on our driveway, wouldn't see me. I ducked down, and the interpreter covered me with her coat. They left me hunched over like that while they got out.

As I sat there, I thought about Meredith. She was quiet and kept to herself at home, but she was smart, cheerful, generous.

I still couldn't believe that she was gone. I was overwhelmed by the enormity, the finality of her death. I wondered how her family was coping with the news. Meredith had told me that her mother had health issues, and I hoped her daughter's sudden, shocking death hadn't triggered an episode. I felt sorry for Meredith's sister. *What would I do if something were to happen to one of mine?*

When the police finally came to get me, I saw that the entrance to our apartment was blocked off with yellow police tape. Instead of going in, the police had me show them from the outside what I'd noticed about Filomena's window, asking whether the shutters were opened or closed when Raffaele and I had come home. They wanted details about how we lived. Did we usually lock the gate to our driveway? What about the faulty lock on the front door? Did anyone else have a key? Were there outside lights on at night? Did Meredith often stay there alone? Did we have frequent visitors?

Then the police led me around back to the downstairs apartment. The glass in the guys' front door had been shattered and lay everywhere. I gasped, thinking someone had since broken in there, too. The police said, "No, no, no. We kicked it in ourselves." They handed me protective booties and gloves. After I slipped them on, I sang out, "Ta-dah," and thrust out my arms like the lead in a musical.

It was an odd setting for anything lighthearted, but having just been reprimanded for complaining, I wanted to be friendly and show that I was cooperating. I hoped to ease the tension for myself, because this was so surreal and terrifying. Instead of smiling, they looked at me with scorn. I kept trying to recalibrate my actions, my attitude, my answers, to get along, but I couldn't seem to make things better no matter what I did. I wasn't sure why.

I followed behind them in silence. We stopped first at Stefano's room. The comforter on his bed was crumpled up and stained with blood. I took another sharp breath. They said, "Do you see anything not normal?"

It seemed a bizarre thing to ask. I said, "Yes, there are blood-stains." The sight of it made my heart and mind race. I was trying to piece together what I'd seen. The agonizing thought that maybe Meredith had been attacked downstairs and chased back into our apartment before she was killed struck me like a physical blow. I kept thinking about how utterly terrified she must have been. I wanted to know what she had been through in her final moments, but at the same time I couldn't bear to go there.

I didn't think I could take any more surprises, but they kept coming. Next, the police opened up a closet to reveal five thriving marijuana plants. "Does this look familiar?" they asked.

"No," I said. Despite my earlier lie about not smoking in our house, I was now telling the truth. I was stunned that the guys were growing a mini-plantation of pot. I couldn't believe I had talked to them every day since I'd moved in six weeks earlier and they'd never mentioned it. I said, "I don't really hang out down here a lot."

Next we went to the room that Marco and Giacomo shared. There was no blood—or contraband plants. While we stood there, the detectives started asking me pointed questions about Giacomo and Meredith. How long had they been together? Did she like anal sex? Did she use Vaseline?

"For her lips," I said. When I'd first gotten to town, Meredith and I had hunted around at different grocery stores until we found a tiny tub of Vaseline.

Giacomo and Meredith had definitely had sex, but I certainly didn't know which positions they'd tried. Meredith didn't talk about her sex life in detail. The most she'd done was ask me once if she could have a couple of the condoms I kept stashed with Brett's still-unused gift, the bunny vibrator, in my see-through beauty case in the bathroom Meredith and I shared.

I couldn't understand why the police were asking me about anal sex. It disturbed me. Were they hinting that Meredith had been raped? What other unthinkably hideous things had happened to her?

After that, I was taken back to the car and left alone. I felt as though I'd been emotionally thrashed, and I lay in the backseat staring blankly at the floor. The interpreter came up to the window and asked if I was all right.

"No," I said. "I'm confused and tired, and I can't help imagining all the horrible things Meredith must have gone through."

Back at the *questura*, I had to repeat for the record everything I'd been asked about at the villa. It was a tedious process at the end of a difficult day.

Finally, at around 7 P.M., I was allowed to call Raffaele to pick me up. While I was waiting for him, Aunt Dolly phoned. "Did you ask the police if you can leave Perugia? If you can come to Germany?" she asked.

"Yeah, and they said no, that I'd have to wait until they heard from the magistrate in three days. Whatever that means."

"If they really want to question you, they can do it from Germany," Dolly said. "Maybe we should find a lawyer for you so they don't say you have to stay there forever."

"Yeah. I'll wait to see what they say in three days."

"Okay," Dolly said. "Let's wait for Tuesday. Then, if they don't let you know when you can leave, tell them you have to contact the American embassy to get a lawyer. Without a doubt the embassy will help you, Amanda."

As I walked outside the *questura*, I saw the guys from downstairs coming in. After we said hello, I wavered for a moment over the police's order that I never talk about what I saw. "I was at your apartment today and you should know that your comforter was splotched with blood, Stefano. It made me wonder if Meredith was down there before she died. It was awful."

"Yeah," Stefano, said. "I hope that was from our cat and not Meredith." Stefano, Giacomo, and Marco exchanged anxious looks.

Just then, Raffaele drove up and I said good-bye to the guys. Raffaele took me to a small boutique downtown called Bubble, next door to a luxury lingerie shop. Pulsating with music, Bubble catered to students, offering trendy, cheaply made clothing, the kind that's not meant to outlast a season. I tried on a few things but decided to wait until my mom got to town to replace my staples, which were locked in the crime scene. I settled on one necessity, grabbing a pair of cotton bikini briefs in my size from a display rack near the cash register. In the long run it probably would have been better if I'd chosen a more sedate color than red. I didn't give it another thought, but it turned out that what was insignificant to me was a big deal to other people. Standing at the cash register as he paid, Raffaele hugged me and gave me a few kisses—our lingua franca in a scary, sad time. A few weeks later, the press would report that I bought "a saucy G-string" and that Raffaele brazenly announced: "I'm going to take you home so we can have wild sex together."

The main event of the night was getting together with Laura and Filomena. I was living inside a vacuum—I was staying with

Raffaele, I'd spent the day with the police, and I was talking to the disembodied voices of my family. I needed the ballast I knew my flatmates would give me. We sat around their friend's kitchen table and talked while Laura and Filomena chain-smoked. It was good to download with them. We were as close to normal as possible and wasted no time in speculating about what could possibly have happened at the villa. One scenario was that Meredith had come into the house, locked the front door behind her, and then the murderer broke in to find her there. After killing her, he stole her key to unlock the front door and ran, leaving the door swinging open in the frigid November breeze. Another was that Meredith had come home alone to find the killer biding his time, waiting inside for the first person through the door. I told them about the blood in the downstairs apartment and my worries that Meredith had been chased around there.

All these possibilities left me feeling chilled.

"It's strange to me that whoever did this only wrecked one room and didn't steal anything. Why do you think that was?" I asked.

Laura said, "I don't know. We just have to let the police do their job. I'm sure they'll figure it out."

"The police are grilling me endlessly," I said.

Filomena said, "I know it's hard, Amanda. You've just got to be patient. They're fixated on you because you knew Meredith better than we did."

Laura and Filomena were each consulting a lawyer about how to get out of the lease. No doubt their lawyers were also counseling them on other things, such as how to deal with the police and on our pot-smoking habit, but they didn't mention any of that.

"Are you okay living with Raffaele? How's it going?" Laura asked. "Filomena and I are thinking about sharing another place."

"Would you guys mind if I live with you again?"

Laura said, "Of course you can live with us."

They both hugged me.

"Don't worry. Everything will be okay," Filomena said.

We'd heard that Meredith's parents were coming to Perugia, and we decided to meet them together. "I'm sure they'd like to hear how kind Meredith was to us," Filomena said.

It was after midnight when Raffaele and I finally went back to his apartment. I stayed up surfing the Internet on his computer, looking for articles about the case. As many answers as the police had demanded of me, they weren't giving up much information. Then I wrote a long e-mail, which I sent to everyone at home, explaining what had happened since I'd gone back to the villa on Friday morning. I wrote it quickly, without a lot of thought, and sent it at 3:45 A.M.

It was another night of fretful sleep.

Chapter 9

November 4, 2007, Day Three

If I'd thought about it at all in the days after Meredith's body was discovered, I would have said that my innocence was so obvious no one could possibly miss it. By assuming that I didn't need safeguards, I became vulnerable.

Had I seen a news item that morning in *The Mail on Sunday*, a London tabloid, it might have shifted everything for me. The article said the Italian police were investigating the possibility that the murderer was a woman—someone whom Meredith had known well. " 'We are questioning her female housemates as well as her friends,' a senior police detective said."

Or I might simply have thought: *It's not Laura, it's not Filomena, it's not me. Whom could they possibly be thinking of?*

The first thing the police wanted at the *questura* that day was a list of every man who'd ever been inside our villa. Just as they'd done the day before, they prodded me with questions like "Who do you think would do this? Whom do you know who disliked Meredith? Was there anyone who might have had a motive?"

That afternoon, back in the waiting room by myself, I slouched down in one of the hard plastic chairs that had molded to my body over the past forty-eight hours. Exhaustion and emotion had overtaken me.

In quiet moments like this, as in the squad car the day before, my thoughts went straight to Meredith and the torture she'd been put through. I tried to imagine over and over how she might have died, what might have happened, and why. I replayed memories of our hours spent on the terrace talking, our walks around town, the people we'd met, the last time I'd seen her.

Either Meredith's murder was completely arbitrary or, worse, irrationally committed by a psychopath who had targeted our villa as Chris had suggested. The hardest question I put to myself was: *What if I'd been home that night? Could I have saved Meredith? Would she somehow still be alive?*

I was lost in these black thoughts when the interpreter walked by, looked at me, and said, "Oh my God. Are you okay?"

I thought, *I'm numb. I can't sleep and I've just started my period. Every part of my brain is screaming to make sense of this.* But I said, "I'm just tired."

"You're pale," she said. "Maybe a cappuccino would help. Come with me."

When we got back upstairs from the vending machines, Raffaele was waiting for me. He'd stopped by to see if I could leave. The police said no, but they let us visit for a few minutes, in the same office where I'd been questioned the first night. I didn't know that the room, like our cell phones, was bugged.

We stood together, talking quietly about nothing. I leaned against him, glad for his company. He kissed me.

Just then, Rita Ficarra, the police officer who'd said I couldn't leave Perugia, walked by. She turned around and gave us a piercing stare. "What you're doing is completely inappropriate," she hissed. "You need to stop this instant."

I was taken aback. It's not like we were making out. What could she possibly think was improper about a few tender hugs and kisses? Raffaele was being compassionate, not passionate—giving me the reassurance I needed. But we were offending her.

Raffaele was the main reason I was able to keep myself somewhat together in those days. I'd known him for such a short time, and he had met Meredith just twice. Who would have blamed him if he hadn't stuck around? Besides giving me a place to stay, he had been patient and kind. He'd dedicated himself to my safety and comfort—driving me to and from the police station, making sure I ate, curling around me at night so I'd feel protected. I had put him on the phone with Mom, Dad, Chris, and Dolly to reassure them. He made sure I was never alone.

I was in touch with Laura and Filomena, but they were busy trying to patch their own lives back together. They had their own friends, and their families were close by. Until my mom arrived on Tuesday morning, Raffaele was all I had.

But as much as he was helping me, we were careening to a bad end together. Whether it was kissing outside the house while Meredith lay inside dead, or whispering, joking, and making faces in the *questura*, our behavior had aroused suspicion. I was oblivious to it, but apparently once the police thought we were guilty, it colored everything.

At Ficarra's insistence, Raffaele and I stepped apart. I stayed in that office after he left, and I was still sitting there when Laura and Filomena were brought in. We kissed on both cheeks and sat down. "They're treating me like a criminal," I said melodramatically. "They keep asking the same questions over and over, like I'm not telling the truth. I don't know why. I'm not lying."

The police took all three of us back to the villa, with Laura and Filomena riding in the backseat of one squad car and the interpreter and me in another.

We ducked under the yellow police tape that blocked off the front door and put on protective blue shoe covers. I hadn't been back in our apartment since Meredith's body was discovered and the Postal Police had ordered us outside. Tingling with fear, I never thought to reprise my "ta-dah" from the day before. My heart was bolting out of my chest as I walked inside, where it seemed as if every inanimate household object—the bowl on the kitchen counter, the couch in the living room—were witnesses to Meredith's death. I tried not to touch anything.

The police told me to go into my room, and they watched while I did. "Is anything missing?" they asked.

"Everything looks okay," I said, my voice small and quavering. I felt like a kid who's terrified to go down the hall in the dark. Distraught, I forgot to check if my own rent money was still in the drawer of my desk.

"Now come back to the kitchen."

I did.

"Open the bottom drawer and look through the knives. Do you see any missing?"

This is where we kept our overflow utensils, the ones we almost never needed. When I pulled open the drawer, stainless steel gleamed up at me. "I don't know if there's one missing or not," I said, trembling. "We don't really use these."

I reached in, pushed a few knives around, and then stood up helplessly. I knew the assortment in the drawer might include the murder weapon—that they were asking me to pick out what might have been used to slash Meredith's throat. Panic engulfed me.

I don't know how long I stood there, arms limp at my sides. I started crying. Someone led me to the couch. "Do you need a doctor?" the interpreter asked.

"No," I whimpered, my chest heaving. I couldn't speak coherently enough between the sobs to explain. I could only think, *I need to get away from here.* I felt the way Filomena must have felt when she looked into Meredith's room two days before. I didn't have to see the blood, the body, the naked foot, to fully imagine the horror.

I sat there hyperventilating, gulping in air until my panic finally ebbed.

Before we left, they took me to Filomena's room to see if I thought it looked the same as when I'd discovered the break-in. It was hard to tell. It had been a mess then and it was still a mess—and I knew that people, including Filomena, had been through it.

"What did you mean on Friday when you said there were feces in the toilet the first time you looked and that they were gone the next?" they asked.

I told them what I'd seen.

I shuddered when I saw that Meredith's closed door was barricaded behind police tape. *Meredith,* I thought.

When we got back to the *questura,* just before 7 P.M., Raffaele was waiting for me with a pizza. I was wolfing it down, sitting in the same office as before, when my phone rang. Mom was leaving in the morning to start hopscotching her way from Seattle to Rome. There were no direct flights. Even though I'd first told her not to come, now I couldn't wait for her to wrap her arms around me at the train station on Tuesday morning.

The next phone call was from Dolly, the family member closest to me by several thousand miles and my de facto emergency contact. Six months ago, my parents' biggest worry about my year abroad was that I might get sick. We didn't have a plan to cover what I should do if my housemate was murdered. Dolly had a newborn and couldn't easily travel to Perugia, but she was checking in with me regularly. Just before we hung up, she gave me a second warning that I should by then have realized myself. "Amanda," she said, "you should call the American embassy in Rome. Just fill them in on what's been happening. It would be a good idea to have it on the record, you know, just in case."

I didn't understand what she was getting at. I thought, *Just in case what?*

I was naïve, in over my head, and with an innate stubborn tendency to see only what I wanted. Above all, I was innocent. There were so many what-ifs that I never even began to contemplate. What if I hadn't thrown the bunny vibrator in my clear makeup case for anyone to see? What if I hadn't gone on a campaign to have casual sex? What if Raffaele and I hadn't been so immature? What if I'd flown home to Seattle right after the murder, or to Hamburg? What if I'd asked my mom to come immediately to help me? What if I had taken Dolly's advice? What if I'd gotten a lawyer?

Chapter 10

November 5, 2007, Day Four

P olice officer Rita Ficarra slapped her palm against the back of my head, but the shock of the blow, even more than the force, left me dazed. I hadn't expected to be slapped. I was turning around to yell, "Stop!"—my mouth half-way open—but before I even realized what had happened, I felt another whack, this one above my ear. She was right next to me, leaning over me, her voice as hard as her hand had been. "Stop lying, stop lying," she insisted.

Stunned, I cried out, "Why are you hitting me?"

"To get your attention," she said.

I have no idea how many cops were stuffed into the cramped, narrow room. Sometimes there were two, sometimes eight—police coming in and going out, always closing the door behind them. They loomed over me, each yelling the same thing: "You need to remember. You're lying. Stop lying!"

"I'm telling the truth," I insisted. "I'm not lying." I felt like I was suffocating. There was no way out. And still they kept yelling, insinuating.

The authorities I trusted thought I was a liar. But I wasn't lying. I was using the little energy I still had to show them I was telling the truth. Yet I couldn't get them to believe me.

We weren't even close to being on equal planes. I was twenty, and I barely spoke their language. Not only did they know the law, but it was their job to manipulate people, to get "criminals" to admit they'd done something wrong by bullying, by intimidation, by humiliation. They try to scare people, to coerce them, to make them frantic. That's what they do. I was in their interrogation room. I was surrounded by police officers. I was alone.

No one read me my rights. I had no idea that I could remain silent. I was sure you had to prove your innocence by talking. If you didn't, it must mean you were hiding something.

I began to trust them even more than I trusted myself. So much pressure was being exerted on me that I couldn't think through what was happening. I was losing my sense of reality. I would have believed, and said, anything to end the torment I was in.

———————

That Monday morning, Meredith's autopsy report was splashed across the British tabloids depicting a merciless, hellish end to her life. The fatal stabbing, the coroner said, had been done with a pocketknife, and skin and hair found beneath Meredith's fingernails showed she was locked in a vicious to-the-death struggle with her killer. Mysteriously, news accounts reported that something in the same report had made the police bring Filomena, Laura, and me back to the villa. To this day I don't know what it was.

There was evidence that Meredith had been penetrated, but none that proved there had been an actual rape. But other clues that would lead the police to the murderer had been left behind. There was a bloody handprint smeared on the wall and a bloody shoeprint on the floor. A blood-soaked handkerchief was lying in the street nearby. As the stories mounted, I was the only one of Meredith's three housemates being mentioned consistently by name: "Amanda Knox, an American," "Amanda Knox, fellow exchange student," "Amanda Knox, Meredith's American flat-mate." It was all going horribly wrong.

But by that time I wasn't paying attention to the news.

I was desperate to get back to my regular routine, an almost impossible quest given that any minute I expected the police to call again. I didn't have a place of my own to live or clean clothes to wear. But trying to be adult in an unmanageable situation, I borrowed Raffaele's sweatpants and walked nervously to my 9 A.M. grammar class. It was the first time since Meredith's body was found that I'd been out alone.

Class wasn't as normal as I would have liked. Just before we began the day's lesson, a classmate raised her hand and asked, "Can we talk about the murder that happened over the week-end?"

I knew I hadn't been singled out, but that's the way it felt. I said, "Can we not? She was my housemate, and the police have asked me not to say anything." The other students murmured vague sympathies, but the attention put me even more on edge.

When my phone rang I drew in my breath, exhaling only after I realized it was Dolly. "Have you reached the American embassy?" she asked.

"No," I said, stepping into the hall. "I haven't had time, but I'll try to figure it out. I'm back in class."

In truth, I hadn't even thought about calling the embassy.

As with everyone who'd phoned, I wanted Dolly to believe that I had my life under control. I was still trying to believe it myself.

In retrospect I understand that Dolly had a hunch I was headed for a train wreck—that in keeping me awake, calling me back in, the police were interested in me as more than just a "person informed of the facts." I didn't see these things as I should have, as foreshadowing, or that Dolly's advice was now my last chance to alter the course of coming events. I just viewed her suggestions as moral support, like other calls I was getting from my family and friends.

She said, "You're a strong girl. I love you. Your mom's going to be there tomorrow, so stay tough."

When class ended I headed back toward Raffaele's apartment. As I walked through Piazza Grimana, I saw Patrick standing in a crowd of students and journalists in front of the University for Foreigners administration building. He kissed me hello on both cheeks. "Do you want to talk to some BBC reporters?" he asked. "They're looking for English-speaking students to interview."

I said, "I can't. The police have told me not to talk to anyone about the case."

"Oh, I'm sorry, I didn't mean to put you in a difficult position," he said.

"That's okay. But Patrick . . ." I hesitated. "I've needed to call you. I don't think I can work at Le Chic anymore. I'm too afraid to go out by myself at night now. I keep looking behind me to see if I'm being followed. And I feel like someone is lurking behind every building, watching me."

"No problem. I completely understand. Don't worry about it."

"Thank you."

We kissed again on each cheek. "*Ciao,*" I said.

That afternoon at Raffaele's, I got a text from one of Meredith's friends—a student from Poland—telling me about a candlelight memorial service for Meredith that night. Everyone was supposed to meet downtown, on Corso Vannucci, at 8 P.M. and walk in a procession to the Duomo. I kept wondering about what I should do. I wanted to be there but couldn't decide if it was a good idea for me to go to such a public event. I was sure the people I ran into would ask me what I knew about the murder. In the end my decision was made for me—Raffaele had somewhere else to be, and I wouldn't have considered going alone. It didn't occur to me that people would later read my absence as another indication of guilt.

At around 9 P.M. Raffaele and I went to a neighbor's apartment for a late dinner. Miserable and unable to sit still, I plucked absentmindedly at his friend's ukulele, propped on a shelf in the living room. At about ten o'clock, while we were eating, Raffaele's phone rang. "*Pronto*," Raffaele said, picking up.

It was the police saying they needed him to come to the *questura* immediately. Raffaele and I had the same thought: *This late? Not again.*

Raffaele said, "We're just eating dinner. Would you mind if I finished first?"

That was a bad idea, too.

While we cleared the table, Raffaele and I chatted quickly about what I should do while he was at the police station. I was terrified to be alone, even at his place, and uneasy about hanging out with someone I didn't know. I could quickly organize myself to stay overnight with Laura or Filomena, but that seemed so complicated—and unnecessary. Tomorrow, when my mom arrived, this wouldn't be a question we'd have to discuss.

"I'm sure it's going to be quick," Raffaele said.

I said, "I'll just come with you."

Did the police know I'd show up, or were they purposefully separating Raffaele and me? When we got there they said I couldn't come inside, that I'd have to wait for Raffaele in the car. I begged them to change their minds. I said, "I'm afraid to be by myself in the dark."

They gave me a chair outside the waiting room, by the elevator. I'd been doing drills in my grammar workbook for a few minutes when a silver-haired police officer—I never learned his name—came and sat next to me. He said, "As long as you're here, do you mind if I ask you some questions?"

I was still clueless, still thinking I was helping the police, still unable or unwilling to recognize that I was a suspect. But as the next hours unfolded, I slowly came to understand that the police were trying to get something out of me, that they wouldn't stop until they had it.

To the unnamed police officer, I said, "Okay, but I've told you everything I know. I don't know what else to say."

"Why don't you keep talking about the people who've been in your house—especially men?" he suggested.

I'd done this so many times in the *questura* I felt as if I could dial it in. And finally someone there seemed nice. "Okay," I said, starting in. "There are the guys who live downstairs."

As I was running through the list of male callers at No. 7, Via della Pergola, I suddenly remembered Rudy Guede for the first time. I'd met him only briefly. I said, "Oh, and there's this guy—I don't know his name or his number—all I know is that he plays basketball with the guys downstairs. They introduced Meredith and me to him in Piazza IV Novembre. We all walked to the villa together, and then Meredith and I went to their apartment for a few minutes."

While we talked, I got up to stretch. I'd been sitting hunched over a long time. I touched my toes, flexed my quads, extended my arms overhead.

He said, "You seem really flexible."

I replied, "I used to do a lot of yoga."

He said, "Can you show me? What else can you do?"

I took a few steps toward the elevator and did a split. It felt good to know I still could.

While I was on the floor, legs splayed, the elevator doors opened. Rita Ficarra, the cop who had reprimanded Raffaele and me about kissing the day before, stepped out.

"What are you doing?" she demanded, her voice full of contempt.

I stood up and returned to my chair. "Waiting," I said.

The silver-haired officer said, "I was just asking Amanda some questions."

Ficarra said, "If that's the case, we need to put it on the record."

She led me through the waiting room and into the same office with the two desks where I'd spent so much time. As we were walking, she looked at me, narrowing her eyes. "You said you guys don't smoke marijuana. Are you sure you're being honest?"

"I'm really sorry I said that." I grimaced. "I was afraid to tell you that all of us smoked marijuana occasionally, including Meredith. We'd sometimes pass a joint around when we were chilling out with the guys or with Filomena and Laura. But Meredith and I never bought any pot; we didn't know any drug dealers."

She shut the door and signaled for me to sit down on a metal folding chair, taking the seat across the desk from me. The silver-haired officer pulled up a chair next to me, effectively cutting the room in half. The light was bright. The walls were blank. I had nowhere to look but at the police. They said, "We're going to call in an interpreter."

While we waited for the interpreter to arrive, they said, "Tell us more about the last time you saw Meredith."

I did.

Then they said, "Okay, minute by minute, we want you to tell us what happened."

I still thought they were using me to find out more information about Meredith—her habits, whom she knew, who could possibly have had a motive to kill her. I started trying to describe the exact time I saw Meredith leave the house. I said, "I think it was around two P.M.—one or two. I'm not sure which. I don't wear a watch, and the time didn't matter—it was a holiday. But I know it was after lunch."

Then the questions shifted. They asked, "When did you leave your house?"

At first, when they started questioning me about what I did, I thought they were just trying to test whether I was telling the truth—maybe because I'd lied about our marijuana use.

I said, "Before dinner—four-ish maybe."

They said, "Are you sure it was four-ish? Was it four o'clock or five o'clock? You didn't see the time?"

"No. Then we went to Raffaele's place."

"How long it did it take you to get there?"

"I don't know—a couple of minutes. He doesn't live far away."

"What happened then?"

"Nothing happened. We had dinner; we watched a movie; we smoked a joint; we had sex; we went to bed."

"Are you positive? Nothing else?"

"Well, I got a text message from my boss telling me I didn't have to work that night."

"What time did that happen?"

"I think around eight P.M.—maybe. Maybe it was before then."
I was thinking, *It had to be before I'd normally go to work.* "Maybe
seven or eight?"

That wasn't good enough for them.

They kept asking me for exact times, and because I couldn't
remember what had happened from 7 P.M. to 8 P.M. and 8 P.M. to
9 P.M. they made it seem as if my memory were wrong. I started
second-guessing myself. Raffaele and I had done some varia-
tion of watching a movie, cooking dinner, reading *Harry Potter*,
smoking a joint, and having sex every night for the past week.
Suddenly it all ran together so that I couldn't remember what
time we'd done what on Thursday, November 1. I kept saying,
"I'm sorry, I'm sorry."

I was afraid to say that I didn't know the difference between
7 P.M. or 8 P.M., and I was beginning to feel panicky because they
were demanding that I know. My heart was hammering, my
thoughts were scrambled, and the pressure on the sides of my head
made it feel as if my skull were going to split apart. I couldn't think.
Suddenly, in trying to distinguish between this time or that time,
this sequence of events or that one, I started forgetting everything.
My mind was spinning. I felt as if I were going totally blank.

"Which was it?"

I took a deep breath. "I don't remember."

Ficarra thrust her hand out aggressively and insisted, "Let me
see your cell phone."

I handed it to her. As they looked through it, they kept pounding me with questions. "What movie did you watch?"

"*Amélie.*"

"How long is that movie?"

"I don't know."

"Did you watch it all the way through?"

"Well, we paused it at some point, because we noticed that the sink was leaking."

"But you said you'd had dinner before that."

"I guess you're right. I think the sink leaked before we watched the movie, but then I remember pausing it."

"Why did you pause it?"

"I don't remember."

"Why? Why? What time?"

"I don't remember!" I said it forcefully, trying to shake them off, but it didn't work. They were peppering me relentlessly. The questions seemed simple, but I didn't have the answers. And the more they asked, the more I lost my bearings. I was getting hot, looking around for air. I was having my period, and I could feel myself bleeding into my underwear. "I need to use the bathroom," I said. "I have a feminine issue."

"Not right now," they said. "Did you pause the movie before dinner or after?"

"I think it was after we had dinner, but now that I think about it, it seemed pretty late when we had dinner."

"Why can't you remember? Did you have dinner before or after the movie?"

"You're freaking me out," I yelled. "I can't think when I'm freaked out. It just seemed late when we ate."

By now their tone was shrill. "Why can't you just tell us? Why can't you remember?"

I could tell they thought I was lying. I said, "I'm sorry, it's hard to remember, and I'm really tired. There are some nights we had dinner earlier and some nights later. It seemed late to me, but I don't remember what time it was."

I was exhausted. I hadn't slept but a few hours in the past four days, and the back-and-forth to the police station—on top of the shock I felt over Meredith—had left me empty. I didn't know I could say, "We need to stop, because I'm too tired." I was ashamed that I couldn't answer their questions, that I was failing. I didn't know what to do to make it better. I wanted so badly to appease them so they would go away.

The interpreter, a woman in her forties, arrived at about 12:30 A.M. It's inconceivable to me now that all the questioning up to that point had been in Italian. For a couple of hours I'd done my best to hang in there, to grasp what they were saying. I kept saying, "Okay, I understand." I was always mortified when I had to admit that my Italian wasn't up to speed.

The truth is that although I could guess what they meant, this was another case of my false bravado. By that time, my Italian was fine for exchanging pleasantries over a cup of tea. But in no credible way was it strong enough, after only six weeks in Italy, for me to be defending myself against accusations of murder.

The interpreter sat down behind me. She was irritated and impatient, as if I were the one who had rousted her from bed in the middle of the night.

The silver-haired cop and Ficarra were in the tiny room almost nonstop. When they left, it wasn't for long, and other cops came in to take their place. Sometimes a crowd of people closed in on me. The room was becoming uninhabitable for me. I really had to use the bathroom, to take care of my period, but now I was too afraid to ask.

Just then a cop—Monica Napoleoni, who had been so abrupt with me about the poop and the mop at the villa—opened the door. "Raffaele says you left his apartment on Thursday night," she said almost gleefully. "He says that you asked him to lie for you. He's taken away your alibi."

My jaw dropped. I was dumbfounded, devastated. *What?* I couldn't believe that Raffaele, the one person in Italy whom I'd trusted completely, had turned against me. How could he say that when it wasn't true? We'd been together all night. Now it was just me against the police, my word against theirs. I had nothing left.

"Where did you go? Who did you text?" Ficarra asked, sneering at me.

"I don't remember texting anyone."

They grabbed my cell phone up off the desk and scrolled quickly through its history.

"You need to stop lying. You texted Patrick. Who's Patrick?"

"My boss at Le Chic."

"What about his text message? What time did you receive that?"

"I don't know. You have my phone," I said defiantly, trying to combat hostility with hostility. I didn't remember that I'd deleted Patrick's message.

They said, "Why did you delete Patrick's message? The text you have says you were going to meet Patrick."

"What message?" I asked, bewildered. I didn't remember texting Patrick a return message.

"This one!" said an officer, thrusting the phone in my face and withdrawing it before I could even look. "Stop lying! Who's Patrick? What's he like?"

"He's about this tall," I said, gesturing, "with braids."

"Did he know Meredith?"

"Yes, she came to the bar."

"Did he like her?"

Me, age four.

Me at age thirteen,
the year my
teammates dubbed
me Foxy Knoxy
for the way I moved
the soccer ball
down the field.

My sister Deanna and me on the train en route to Perugia, August 2007.

My first afternoon at No. 7, Via della Pergola, with my future Italian roommates. *From left:* Filomena Romanelli, me, Laura Mezzetti.

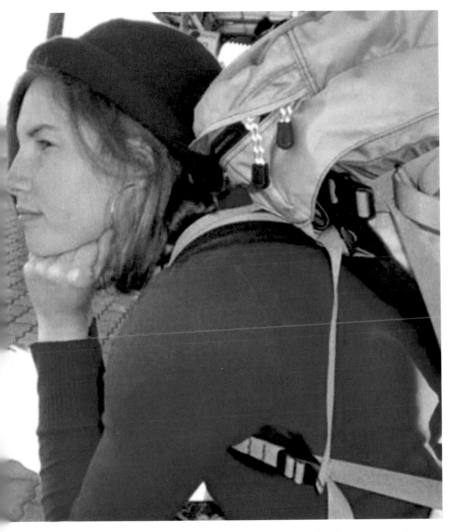

At the train station in Perugia in September 2007, embarking on my long-awaited year abroad.

Meredith Kercher, our fourth roommate at No. 7, Via della Pergola, was twenty-one and an exchange student from outside London.
(Press Association via AP Images)

Raffaele Sollecito, my boyfriend of one week,
whom I met at a classical music performance.

The kitchen/living area at No. 7, Via della Pergola, where we often relaxed together with our downstairs neighbors.
(Iberpress/Barcroft Media)

Diya "Patrick" Lumumba, my boss at the bar Le Chic.
(Tiziana Fabi/AFP/Getty Images)

At the villa on Saturday, November 3, I'm explaining to detectives what I'd seen upon arriving home the day before, when Meredith's body was found. I didn't realize that the police had already considered me a murder suspect.
(© Pietro Crocchioni/epa/Corbis)

Police officer Lorena Zugarini (*left*) and Rita Ficarra,
my chief interrogator (*right*), lead me to a squad car after my arrest.
(*© Pietro Crocchioni/epa/Corbis*)

"Yes, he liked Meredith. He was nice to her, and they got along."

"Did he think Meredith was pretty?"

"Well, Meredith was pretty. I'm sure he thought she was pretty."

"When did you leave to meet Patrick?"

"I didn't meet Patrick. I stayed in."

"No, you didn't. This message says you were going to meet him."

"No. No, it doesn't."

They read the message aloud: *"Certo ci vediamo più tardi buona serata!"*—"Okay, see you later, have a good evening!"

"That means 'we're going to see each other,'" they said, translating the *Ci vediamo* for me. "You said, 'See you later.' Why did you go see him?"

"I didn't see him!" I shouted. "In English, 'see you later' means good-bye. It doesn't mean we're going to see each other *now*. It means see you *eventually*."

In my beginner's Italian, I had had no idea that I'd used the wrong phrase in my text to Patrick—the one that means you're going to see someone. I'd merely translated it literally from the English.

The interpreter balked: "You're a liar."

"No, I'm not! I never left Raffaele's apartment."

The detectives said, "You did leave. Raffaele said you left. You said you were meeting Patrick."

How could I make them believe that I'd been at Raffaele's all night? My protests seemed so flimsy, especially when they ganged up on me. I couldn't make them believe anything.

I said, "I didn't leave."

"Who did you meet up with? Who are you protecting? Why are you lying? Who's this person? Who's Patrick?"

The questions wouldn't stop. I couldn't think. And even when it didn't seem possible, the pressure kept building.

I said, "Patrick is my boss."

The interpreter offered a solution, "Once, when I had an accident, I didn't remember it. I had a broken leg and it was traumatizing and I woke up afterward and didn't remember it. Maybe you just don't remember. Maybe that's why you can't remember times really well."

For a moment, she sounded almost kind.

But I said, "No, I'm not traumatized."

Another cop picked up the same language. He said, "Maybe you're traumatized by what you saw. Maybe you don't remember."

Everyone was yelling, and I was yelling back. I shouted, "I don't understand what the fuck is happening right now!"

A beefy cop with a crew cut thought I'd said, "Fuck you," and he yelled, "Fuck you!" back.

They pushed my cell phone, with the message to Patrick, in my face and screamed, "You're lying. You sent a message to Patrick. Who's Patrick?"

That's when Ficarra slapped me on my head.

"Why are you hitting me?" I cried.

"To get your attention," she said.

"I'm trying to help," I said. "I'm trying to help, I'm desperately trying to help."

The pressure was greater than just being closed in a room. It was about being yelled at relentlessly by people I trusted completely, by people I'd been taught to respect. Everything felt bigger, more overwhelming, more suffocating, than it was because these were people whom I thought I was helping and they didn't believe me; they kept telling me I was wrong.

They told me I'd been to our house, that they had evidence to prove it. They told me I'd left Raffaele's. Raffaele himself had said so. They told me I'd been traumatized and had amnesia. I hadn't slept in days. They wouldn't let me leave the room or give me a moment to think. Nothing had substance. Nothing seemed real. I believed them. Their version of reality was taking over. I felt confused, frantic, and there was no escape.

People were shouting at me. "Maybe you just don't remember what happened. Try to think. Try to think. Who did you meet? Who did you meet? You need to help us. Tell us!"

A cop boomed, "You're going to go to prison for thirty years if you don't help us."

The threat hung in the air. I was feeling smaller and smaller, more and more helpless. It was the middle of the night. I was terrified, and I couldn't understand what was happening. I thought they had to be pressuring me for a reason. They had to be telling me the truth. Raffaele had to be telling the truth. I didn't trust my own mind anymore. I believed the police. I could no longer distinguish what was real from what wasn't. I had a moment when I thought I was remembering.

The silver-haired police officer took both of my hands in his. He said, "I really want to help you. I want to save you, but you need to tell me who the murderer is. You need to tell me. You know who the murderer is. You know who killed Meredith."

In that instant, I snapped.

I truly thought I remembered having met somebody. I didn't understand what was happening to me. I didn't understand that

I was about to implicate the wrong person. I didn't understand what was at stake. I didn't think I was making it up. My mind put together incoherent images. The image that came to me was Patrick's face.

I gasped. I said his name. "Patrick—it's Patrick."

I started sobbing uncontrollably. They said, "Who's Patrick? Where is he? Where is he?"

I said, "He's my boss."

"Where did you meet him?"

"I don't remember."

"Yes, you do."

"I don't know—at the basketball court."

"Why did he kill her? Why did he kill her?"

I said, "I don't know."

"Did he have sex with Meredith? Did he go into the room with Meredith?"

"I don't know, I guess so. I'm confused."

They started treating me like someone who'd been taken advantage of. They told me they were helping me, that they were trying to get to the truth. "We're trying to do our best for you."

They were softer, but I was no longer sure of anything—of what was real, of what I feared, of what I imagined.

I wept for a long time.

At 1:45 A.M. they gave me a piece of paper written in Italian and told me to sign it.

On Thursday, November 1, on a day when I normally work, while I was at my boyfriend Raffaele's place, at about 20:30, I received a message on my cell phone from Patrik, who told me the club would remain closed that night because there weren't any customers and therefore I would not have to go to work.

I replied to the message telling him that we'd see each other right away. Then I left the house, saying to my boyfriend that I had to go to work. Given that during the afternoon with Raffaele I had smoked a joint, I felt confused because I do not make frequent use of drugs that strong.

I met Patrick immediately at the basketball court in Piazza Grimana and we went to the house together. I do not remember if Meredith was there or came shortly afterward. I have a hard time remembering those moments but Patrick had sex with Meredith, with whom he was infatuated, but I cannot remember clearly whether he threatened Meredith first. I remember confusedly that he killed her.

As soon as I signed it, they whooped and high fived each other.

Then, a few minutes later, they demanded my sneakers. As soon as I took them off, someone left the room with them.

Eventually they told me the *pubblico ministero* would be coming in. I didn't know this translated as prosecutor, or that this was the magistrate that Rita Ficarra had been referring to a few days earlier when she said they'd have to wait to see what he said, to see if I could go to Germany. I thought the "public minister" was the mayor or someone in a similarly high "public" position in the town and that somehow he would help me.

They said, "You need to talk to the *pubblico ministero* about what you remember."

I told them, "I don't feel like this is remembering. I'm really confused right now." I even told them, "I don't remember this. I can imagine this happening, and I'm not sure if it's a memory or if I'm making this up, but this is what's coming to mind and I don't know. I just don't know."

They said, "Your memories will come back. It's the truth. Just wait and your memories will come back."

The *pubblico ministero* came in.

Before he started questioning me, I said, "Look, I'm really confused, and I don't know what I'm remembering, and it doesn't seem right."

One of the other police officers said, "We'll work through it."

Despite the emotional sieve I'd just been squeezed through, it occurred to me that I was a witness and this was official testimony, that maybe I should have a lawyer. "Do I need a lawyer?" I asked.

He said, "No, no, that will only make it worse. It will make it seem like you don't want to help us."

It was a much more solemn, official affair than my earlier questioning had been, though the *pubblico ministero* was asking me the same questions as before: "What happened? What did you see?"

I said, "I didn't see anything."

"What do you mean you didn't see anything? When did you meet him?"

"I don't know," I said.

"Where did you meet him?"

"I think by the basketball court." I had imagined the basketball court in Piazza Grimana, just across the street from the University for Foreigners.

"I have an image of the basketball court in Piazza Grimana near my house."

"What was he wearing?"

"I don't know."

"Was he wearing a jacket?"

"I think so."

"What color was it?"

"I think it was brown."

"What did he do?"

"I don't know."

"What do you mean you don't know?"

"I'm confused!"

"Are you scared of him?"

"I guess."

I felt as if I were almost in a trance. The *pubblico ministero* led me through the scenario, and I meekly agreed to his suggestions.

"This is what happened, right? You met him?"

"I guess so."

"Where did you meet?"

"I don't know. I guess at the basketball court."

"You went to the house?"

"I guess so."

"Was Meredith in the house?"

"I don't remember."

"Did Patrick go in there?"

"I don't know, I guess so."

"Where were you?"

"I don't know. I guess in the kitchen."

"Did you hear Meredith screaming?"

"I don't know."

"How could you not hear Meredith screaming?"

"I don't know. Maybe I covered my ears. I don't know, I don't know if I'm just imagining this. I'm trying to remember, and you're telling me I need to remember, but I don't know. This doesn't feel right."

He said, "No, remember. Remember what happened."

"I don't know."

At that moment, with the *pubblico ministero* raining questions down on me, I covered my ears so I could drown him out.

He said, "Did you hear her scream?"

I said, "I think so."

My account was written up in Italian and he said, "This is what we wrote down. Sign it."

I want to voluntarily report what happened because I'm deeply disturbed and very frightened of Patrick, the African owner of the pub called "Le Chic" on Alessi Street where I work occasionally. I met him on November 1 at night after I sent a reply to his message with the words "see you later." We soon met about 9 pm at the basketball court in Piazza Grimana. We went to my house on Via della Pergola No. 7. I cannot remember exactly if my friend Meredith was already in the house or if she came after, but I can say that she disappeared into her bedroom with Patrick while I think I stayed in the kitchen. I can't remember how long they were in her bedroom but at one point I heard Meredith screaming and I was scared and covered my ears. I do not remember anything after that. I have a lot of confusion in my head. I do not remember if Meredith screamed or if I heard any thuds because I was in shock, but I could imagine what was going on.

After I signed it, everyone mercifully stopped questioning me, but my mind wouldn't rest. Something didn't feel right. It didn't seem as though I had actually remembered what I said I had. It seemed made up.

In my dull state I thought everything would eventually be okay. I thought I could communicate with people on the outside. My mother was coming that day, and she'd help me figure things out.

I had no more than a shred of memory, but it seemed to hold the truth. I was so afraid of the police, so afraid of sending them in the wrong direction for the wrong person. *What if I've told them wrong? What if I don't have amnesia?*

And what about the "spontaneous declarations," as the police called what I'd signed? These documents didn't take into account that I kept yelling, "I don't know." They didn't say that the police threatened me and yelled at me. None of that is there.

The declarations were in the detectives' words. But now their words were mine, and this shaped everything that followed.

Chapter 11

—————

Morning, November 6, 2007, Day Five

I signed my second "spontaneous declaration" at 5:45 A.M., just as the darkness was beginning to soften outside the small window on the far side of the interrogation room. That was also true on the inside. As soon as I finished crossing the *x* in "Knox," the agonizing torment ended.

The room emptied in a rush. Except for Rita Ficarra, who sat at the wooden desk where she'd been all night, I was alone in the predawn hush.

Just a few more hours and I'll see Mom, I thought. *We'll spend the night in a hotel.*

I asked permission to push two metal folding chairs together, balled myself into the fetal position, and passed out, spent. I probably didn't sleep longer than an hour before doubt pricked me awake. *Oh my God, what if I sent the police in the wrong direction? They'll be looking for the wrong person while the real killer escapes.* I sat up crying, straining to remember what had happened on the night of Meredith's murder. *Had I really met Patrick? Had I even been at the villa? Did I make all that up?* I was too exhausted, too rattled, to think clearly. I was gripped by uncertainty about what I'd said to the police and the *pubblico ministero*. I tried to get Ficarra's attention. "Um, *scusi*," I murmured tentatively. "I'm not sure what I told you is right."

"The memories will come back with time," Ficarra answered mechanically, barely raising her eyes to look at me. "You have to think hard."

It seemed impossible that I could forget seeing a murder. Still, without feeling sure, I thought I should believe her.

I tried to weave the images that had flashed in my mind the night before into a coherent sequence. But my memories—of Patrick, the villa, Meredith's screams—were disjointed, like pieces of different jigsaw puzzles that had ended up in the same box by mistake. They weren't ever meant to fit together. I'd walked by the basketball court near the villa every day. I'd said, "It was Patrick," because I saw his face. I imagined him in his brown jacket because that's what he usually wore. The more I realized how fragmented these images were, the closer I came to understanding that they weren't actual memories.

Suddenly my cell phone, which had been lying on the desk since it was waved in my face, lit up and started ringing. Ficarra ignored this. "Can I please answer it?" I begged. "I'm sure it's my mom; I'm supposed to meet her at the train station. She'll freak out if I don't answer."

"No," Ficarra said. "You cannot have your phone back. Your phone is evidence."

This moment exemplified how the line between Before and After was marked. I'd stopped being in charge of my life.

For the next half hour my phone rang every few minutes,

stopping only while the calls were sent to voice mail. The noise ripped at me, and I began to panic, my body shook. Mom would be sick with worry, wondering what had happened to me, where I was, why I wasn't answering. As a teenager, if I was late checking in, she'd keep trying me until I'd finally pick up, almost always to hear her crying on the other end of the line. I couldn't stand that I was putting her through that now. And now, more than ever, I needed her.

Still, it was a huge relief to know that later, if I had to come back to the *questura*, my mom would come with me. If they didn't need me, I planned to introduce her to Laura and Filomena—and maybe to Meredith's parents, when they arrived.

Finally my phone went silent. I slumped down in the folding chair, as mute as my cell phone.

I was waiting for the police to tell me what they wanted from me next. That had been the pattern at the *questura* for the past four days. There would be a lull, and then they would either question me again or send me home. I willed it to be the latter. I couldn't bear for them to yell at me again.

Around 2 P.M. on Tuesday—it was still the same day, although it felt as if it should be two weeks later—Ficarra took me to the cafeteria. I was starving. After the interrogation was over they brought me a cup of tea, but this was the first food or drink I'd been offered since Raffaele and I had arrived at the *questura* around 10:30 P.M. Monday. With my sneakers confiscated, I trailed her down the stairs wearing only my socks. She turned and said, "Sorry I hit you. I was just trying to help you remember the truth."

I was still too confused to know what the truth was.

I tried to say, "I hope that once this gets sorted out you'll see

I'm on your side." But the way my Italian came out was "I hope you can see I'm your friend."

I was desperate for a sign that everything was okay between us, to be reassured that they still trusted me. I told myself they'd bullied me because they were so stressed, determined to figure out who'd killed Meredith. I had the same feelings. But in re-thinking the night, I decided that the police thought I'd been hiding facts from them, that I'd lied. That's why they were angry with me.

I didn't want them to think I was a bad person. I wanted them to see me as I was—as Amanda Knox, who loved her parents, who did well in school, who respected authority, and whose only brush with the law had been a ticket for violating a noise ordinance during a college party I'd thrown with my housemates in Seattle. I wanted to help the police track down the person who'd murdered my friend.

What I did not know was that the police and I had very different ideas about where I stood. I saw myself as being helpful, someone who, having lived with Meredith, could answer the detectives' questions. I would do that as long as they wanted. But the police saw me as a killer without a conscience. It would be a long time before I figured out that our presumptions were exactly the opposite of each other's.

By the time Ficarra and I got to the cafeteria, lunch was nearly over. I asked for an espresso, and the barista scavenged a few slices of salami and a piece of bread from the slim sandwich makings that were left. When we went back upstairs, a police officer handed me my hiking boots. Someone at the *questura* had gone back to the villa to get them. I'm sure they'd used it as an opportunity to comb through my stuff. Still, I was a lot more worried about what was in my head than on my feet. What had come over me? Why was I so confused? Why had I made those statements, which now seemed less and less like the truth?

I repeated to Ficarra the same things I'd said earlier. "What I described last night doesn't seem like memories. I feel like I imagined the events."

"No, your memories will come back, you'll see," she insisted.

"You don't understand," I protested. "The more I'm remembering, the more I think that what I told you was wrong."

I was sure she was dismissing me because I couldn't explain myself well in Italian. I didn't know that it was because she had other plans for me, that our discussion had ended.

"We need to take you into custody," she said. "Just for a couple of days—for bureaucratic reasons."

Custody? What does that mean? Are they taking me to a safe house? The silver-haired cop had told me during my interrogation that they would protect me if I cooperated, if I told them who the murderer was. *Will my mom be there with me? Can I call her?*

What does "bureaucratic reasons" mean? Does it mean they're just processing my paperwork, my spontaneous declarations?

I had so many questions that I didn't ask aloud. But my main thought was *If I'm going into hiding, I need to make sure the police understand that I'm not sure about Patrick.* I'd caved under the police's questioning. It was my lack of resolve that had created this problem, and I had to fix it.

I needed to say that I had doubts about what I'd signed, to let the police know they couldn't rely on my declarations as the truth. I knew that undoing the cops' work would almost surely mean they'd scream at me all over again. As paralyzing as that thought was, I had to risk it. In naming Patrick, I'd unintentionally misled them. What if they thought I did it on purpose? They'd wasted time on me when they could have been out pursuing the real killer.

"Can I have a piece of paper?" I asked Ficarra. "I need to write down in English what I'm trying to tell you, because you apparently don't understand me right now. You can bring the paper to someone who can tell you what it says in Italian. We can communicate better that way. You're telling me that I'm going to remember when I'm telling you that I *am* remembering, and that I doubt what I said is true."

She handed me a few sheets of paper and a pen. "You'd better write fast," she said. "We have to get going."

If I could make them understand, everything would be okay. I sat down and scrawled four pages that became known as my first *memoriale*:

> This is very strange, I know, but really what happened is as confusing to me as it is to everyone else. I have been told there is hard evidence saying that I was at the place of the murder of my friend when it happened. This, I want to confirm, is something that to me, if asked a few days ago, would be impossible.
>
> I know that Raffaele has placed evidence against me, saying that I was not with him on the night of Meredith's murder, but let me tell you this. In my mind there are things I remember and things that are confused. My account of this story goes as follows, despite the evidence stacked against me:
>
> On Thursday November 1, I saw Meredith the last time at my house when she left around 3 or 4 in the afternoon.

Raffaele was with me at the time. We, Raffaele and I, stayed at my house for a little while longer and around 5 in the evening we left to watch the movie Amelie at his house. After the movie I received a message from Patrik, for whom I work at the pub "Le Chic". He told me in this message that it wasn't necessary for me to come into work for the evening because there was no one at my work.

Now I remember to have also replied with the message: "See you later. Have a good evening!" and this for me does not mean that I wanted to meet him immediately. In particular because I said: "Good evening!" What happened after I know does not match up with what Raffaele was saying, but this is what I remember. I told Raffaele that I didn't have to work and that I could remain at home for the evening. After that I believe we relaxed in his room together, perhaps I checked my email. Perhaps I read or studied or perhaps I made love to Raffaele. In fact, I think I did make love with him.

However, I admit that this period of time is rather strange because I am not quite sure. I smoked marijuana with him and I might even have fallen asleep. These things I am not sure about and I know they are important to the case and to help myself, but in reality, I don't think I did much. One thing I do remember is that I took a shower with Raffaele and this might explain how we passed the time. In truth, I do not remember exactly what day it was, but I do remember that we had a shower and we washed ourselves for a long time. He cleaned my ears, he dried and combed my hair.

One of the things I am sure that definitely happened

the night on which Meredith was murdered was that Raffaele and I ate fairly late, I think around 11 in the evening, although I can't be sure because I didn't look at the clock. After dinner I noticed there was blood on Raffaele's hand, but I was under the impression that it was blood from the fish. After we ate Raffaele washed the dishes but the pipes under his sink broke and water flooded the floor. But because he didn't have a mop I said we could clean it up tomorrow because we (Meredith, Laura, Filomena and I) have a mop at home. I remember it was quite late because we were both very tired (though I can't say the time).

The next thing I remember was waking up the morning of Friday November 2 around 10 am and I took a plastic bag to take back my dirty clothes to go back to my house. It was then that I arrived home alone that I found the door to my house was wide open and this all began. In regards to this "confession" that I made last night, I want to make clear that I'm very doubtful of the verity of my statements because they were made under the pressures of stress, shock and extreme exhaustion. Not only was I told I would be arrested and put in jail for 30 years, but I was also hit in the head when I didn't remember a fact correctly. I understand that the police are under a lot of stress, so I understand the treatment I received.

However, it was under this pressure and after many hours of confusion that my mind came up with these answers. In my mind I saw Patrik in flashes of blurred images. I saw him near the basketball court. I saw him at my front door. I saw myself cowering in the kitchen with my hands over my ears because in my head I could hear Meredith screaming. But I've said this many times so as to make myself clear: these things seem unreal to me, like a dream, and I am unsure if they are real things that happened or are just dreams my head has made to try to answer the questions in my head and the questions I am being asked.

But the truth is, I am unsure about the truth and here's why:

1. The police have told me that they have hard evidence that places me at the house, my house, at the time of Meredith's murder. I don't know what proof they are talking about, but if this is true, it means I am very confused and my dreams must be real.

2. My boyfriend has claimed that I have said things that I know are not true. I KNOW I told him I didn't have to work that night. I remember that moment very clearly. I also NEVER asked him to lie for me. This is absolutely a lie. What I don't understand is why Raffaele, who has always been so caring and gentle with me, would lie about this. What does he have to hide? I don't think he killed Meredith, but I do think he is scared, like me. He walked into a situation that he has never had to be in, and perhaps he is trying to find a way out by disassociating himself with me.

Honestly, I understand because this is a very scary situation. I also know that the police don't believe things of me that I know I can explain, such as:

1. I know the police are confused as to why it took me so long to call someone after I found the door to my house open and blood in the bathroom. The truth is, I wasn't sure what to think, but I definitely didn't think the worst, that

someone was murdered. I thought a lot of things, mainly that perhaps someone got hurt and left quickly to take care of it. I also thought that maybe one of my roommates was having menstral problems and hadn't cleaned up. Perhaps I was in shock, but at the time I didn't know what to think and that's the truth. That is why I talked to Raffaele about it in the morning, because I was worried and wanted advice.

2. I also know that the fact that I can't fully recall the events that I claim took place at Raffaele's home during the time that Meredith was murdered is incriminating. And I stand by my statements that I made last night about events that could have taken place in my home with Patrik, but I want to make very clear that these events seem more unreal to me than what I said before, that I stayed at Raffaele's house.

3. I'm very confused at this time. My head is full of contrasting ideas and I know I can be frustrating to work with for this reason. But I also want to tell the truth as best I can. Everything I have said in regards to my involvement in Meredith's death, even though it is contrasting, are the best truth that I have been able to think.

I'm trying, I really am, because I'm scared for myself. I know I didn't kill Meredith. That's all I know for sure. In these flashbacks that I'm having, I see Patrik as the murderer, but the way the truth feels in my mind, there is no way for me to have known because I don't remember FOR SURE if I was at my house that night. The questions that need answering, at least for how I'm thinking are:

1. Why did Raffaele lie? (or for you) Did Raffaele lie?
2. Why did I think of Patrik?
3. Is the evidence proving my pressance at the time and place of the crime reliable? If so, what does this say about my memory? Is it reliable?
4. Is there any other evidence condemning Patrik or any other person?
5. Who is the REAL murder? This is particularly important because I don't feel I can be used as condemning testimone in this instance.

I have a clearer mind than I've had before, but I'm still missing parts, which I know is bad for me. But this is the truth and this is what I'm thinking at this time. Please don't yell at me because it only makes me more confused, which doesn't help anyone. I understand how serious this situation is, and as such, I want to give you this information as soon and as clearly as possible.

If there are still parts that don't make sense, please ask me. I'm doing the best I can, just like you are. Please believe me at least in that, although I understand if you don't. All I know is that I didn't kill Meredith, and so I have nothing but lies to be afraid of.

I finished writing and handed the pages to Ficarra. I didn't remember the word for "explanation." "This is a present for you"—"*un regalo*," I said.

She said, "What is it—my birthday?"

I felt so much lighter. I knew that I was blameless, and I was sure that was obvious to everyone. We'd just had a misunderstanding. I'd cleared the record.

I was on the police's side, so I was sure they were on mine. I didn't have a glimmer of understanding that I had just made my situation worse. I didn't get that the police saw me as a brutal murderer who had admitted guilt and was now trying to squirm out of a hard-won confession.

My *memoriale* changed nothing. As soon as I gave it to Ficarra, I was taken into the hall right outside the interrogation room, where a big crowd of cops gathered around me. I recognized Pubblico Ministero Giuliano Mignini, who I still believed was the mayor.

An officer stood in front of me as straight as a gun barrel and read me my rights. It was in Italian, only some of which I understood. They handcuffed me. A third person held on to my upper arm. They said, "You're under arrest. We're taking you to prison."

As groggy and mixed-up as I was, those official words startled me. "You're doing what?!" I asked, raising my voice, agitated. I couldn't make sense of this news.

I thought that they were keeping me to protect me. But why would they have to arrest me? And why did they have to take me to prison? I'd imagined that maybe "custody" meant I'd be given a room in the *questura*. That Mom could be there with me.

It's inconceivable to me now that I hardly reacted. It didn't occur to me that I should again ask for a lawyer—or that I needed one. I assumed that once I'd signed my testimony, the moment for a lawyer had passed. I was completely preoccupied with distinguishing between real memories versus whatever I'd imagined. I was lost in my head, trying to remember everything Raffaele and I had done hour by hour, minute by minute, on the night of Meredith's murder so that I could tell the police. I was still replaying my interrogation. I didn't—or couldn't—grasp how much trouble I was in.

If they ever said that I was a murder suspect, I either didn't hear it or didn't understand it. I heard "it's only for a few days," "bureaucratic reasons," and "it's all under control."

"Okay," I said reflexively. I'd fought hard for myself during the night, and I was totally passive now. I had nothing left.

Still, what came next shocked me. After my arrest, I was taken downstairs to a room where, in front of a male doctor, female nurse, and a few female police officers, I was told to strip naked and spread my legs. I was embarrassed because of my nudity, my period—I felt frustrated and helpless. The doctor inspected the outer lips of my vagina and then separated them with his fingers to examine the inner. He measured and photographed my intimate parts. I couldn't understand why they were doing this. I thought, *Why is this happening? What's the purpose of this?*

The doctor and nurse weren't rough with me, but it didn't matter. Being on display, nude, in front of strangers while they discussed me was the most dehumanizing, degrading experience I had ever been through. I didn't protest. I waited silently, feeling violated and angry. In my head I was screaming, *Stop it! Stop it now!*

Next they checked my entire body for cuts and bruises, clawing through my hair to get to my scalp and inspecting the bottoms of my feet. A female police officer pointed out different places to examine and document. I thought, *Why are they measuring the length of my arms and the breadth of my hands? What does it matter how big my feet are?* Later, I realized they were trying to fit the crime to my dimensions. What would Meredith's wounds be like if I'd been the one who stabbed her? *Could* I have stabbed her from my height? They took pictures of anything they thought would be significant.

I pointed out the hickey Raffaele had given me. It had faded to a pinkish tinge on my throat, but I didn't want to appear as if I were hiding anything from them. The police seemed totally uninterested and recorded it perfunctorily. But during my trial the prosecution used it as evidence to fit one of their ever-changing scenarios.

Raffaele. I didn't know what to think of him. How could the person I'd felt so close to have abandoned me? Had he really said, "Amanda left that night" and "Amanda asked me to lie for her?" Or were the police just telling me that? I no longer knew whom I could trust. I felt betrayed and alone.

More than anything, I wanted my mom. She would help me explain what had happened and get me out of this nightmarish experience. *Where is she? How can I reach her? Is she waiting for me at the train station?*

I was finally allowed to get dressed. The police had brought me an airy skirt from the villa with my hiking boots. It seemed like such a ridiculous choice for November that I wadded it up in my purse and put back on Raffaele's clothes, which I'd been wearing before.

I asked to use the bathroom. A female police officer stood in front of the stall with the door open. *Why is she standing here? I can't relax enough to pee, even if she's looking away.* I guessed this unwanted guardian was somehow supposed to keep me safe.

Eventually I put aside my inhibitions long enough to be able to pee. After that they closed the handcuffs back around my wrists. I think they'd left them intentionally loose, but I was so submissive I reported their breach. "Excuse me," I said. "But I can slip my hand out."

They tightened them.

Then they shoved a wool hat down over my eyes. "Duck your head," Ficarra ordered. "Don't look up." She mumbled something about "journalists."

We were standing in a dark foyer. Everything was hushed. My head bent, I was looking at the floor when I suddenly recognized the backs of Raffaele's feet ahead of me. I felt a clenching in my chest. I hadn't seen him since we'd come inside the *questura* together. I had no idea where he'd come from—or why he was walking just steps ahead of me. I so badly wanted to say something, but I knew I shouldn't make a sound.

I just wanted this ordeal to end.

I was consumed by worry for Patrick. I felt that time was running out for him if I didn't remember for sure what had happened the night of Meredith's murder. When I'd said, "It was Patrick," in my interrogation, the police pushed me to tell them where he lived. As soon as I'd mentioned his neighborhood, several officers surrounding me raced out. I figured that they'd gone to question him. I didn't know that it was too late, that they'd staged a middle-of-the-night raid on Patrick's house and arrested him.

Then the doors to the *questura* opened, and I was led outside. No one had told me that what I'd said had been made public. With my head down, it didn't register that there were photographers snapping my picture. Nor could I know that the police would be holding a press conference at which they'd announce, "*Caso chiuso*"—"Case closed." Or that, that evening, news sites would report Raffaele's, Patrick's, and my arrests for "a sexual encounter that went horrifically wrong."

When I look at the pictures of me now—standing in Raffaele's oversize warm-up pants and fleece jacket, a gray wool hat pulled over my eyes—I recall how I followed their directions like a lost, pathetic child. I didn't question, I didn't object, I just put my head down when they told me to and trusted that this would all make sense soon. In that moment, I couldn't see—and it didn't have anything to do with the hat.

I was half-carried, half-pushed from the building, with Ficarra and another person each holding me under an arm. They directed me into a police car, then got in on either side of me. "Duck your head to your knees during the ride," one of the police officers ordered. "Do not try to sit up."

Sirens wailed.

I've since read that the convoy of squad cars drove through Perugia, honking horns in triumph. I only know that we flew along the curving roads in a rush of sound, that we were moving so fast I thought I might get sick in the backseat, that the half-hour trip seemed without end. The officers kept their hands firmly on my back; my eye sockets pressed into my forearms across my knees. The hat pulled down, I was floating, as though I'd escaped from my own body.

Finally our car pulled through the main gate of the Casa Circondariale Capanne di Perugia—not that I knew where we were—and came to a stop inside a dim, cavernous garage. As the doors rumbled closed, I was allowed to sit up. A uniformed prison guard came over, and I tried to catch his eye. I wanted someone, anyone, to look at me and see me for who I was—Amanda Knox, a terrified twenty-year-old girl. He looked through me.

The inner garage door rolled open, and we drove into the prison grounds. My stomach lurched. Concrete walls, ablaze with orange lights and topped with coiled razor wire, stretched up to the night sky in every direction. I felt smaller and more frightened than I'd ever been.

We stopped in front of a single-story building in the center of the complex, where an empty squad car sat. *Raffaele's car?* At a wave from our driver, we entered the building, Ficarra ahead of me, the other officer behind, each gripping one of my arms. Once inside, they let go. "This is where we leave you," they said. One of them leaned in to give me a quick, awkward hug. "Everything's going to be okay. The police will take care of you."

"Thank you," I said. I gave her a last, beseeching look, hoping this meant that finally they knew we were on the same side.

It didn't.

I spent the next 1,427 nights in prison for a crime I did not commit.

Chapter 12

Evening, November 6, 2007, Day Five

O ne guard was trying to flex the thick sole of my hiking boot. The other was shaking her head no.

Of all the things they took from me in my first few minutes as an inmate at Capanne Prison, this loss hit me the hardest. On my nineteenth birthday my stepdad, Chris, had given me his old GPS and taught me how to use it by driving me on a scavenger hunt. We ended up at an outdoor gear store, where I got to pick out my present: the boots I'd coveted for more than a year. I wore them hiking and mountain climbing and paired them with a skirt or dress when I wanted to make an offbeat fashion statement. The boots made me feel invincible— not dangerous, as the guards were implying. Did they think I'd kick someone with the hard, boxy toe? Or try to hang myself with the flimsy laces?

"Do you have other shoes?" the tall, sturdy guard asked me. She had a chiseled jaw and hair that had been dyed reddish pur- ple, like a plum. Her name was Lupa, but prisoners weren't al- lowed to call guards anything but *agente* or *assistente*.

"No, the police took my sneakers," I said. "But they went to my house to get these. Why would they give them to me just to take them away three hours later?"

The other guard, a short, fleshy blonde, continued pawing through my purse/book bag. I later learned the prisoners had nicknamed her Cinema because she spoke in slow motion. "You won't be able to take any of this in with you," she declared flatly.

Everything I needed was in that bag: my wallet, my passport, my journal.

"What about my textbooks?" I asked, pleading. "I have school. I'll be back in class in a few days. I don't want to fall behind."

"When you leave you can request them from the storeroom," Agente Lupa said.

I couldn't believe what was happening. The police told me they would keep me safe, and then they'd just dropped me off here and left. Why would they have done that? They had already confiscated my cell phone and sneakers, and now the prison guards were taking the things that I always kept with me, the things that identified me. Without money, a credit card, my driver's license, my passport, I felt completely vulnerable.

The next orders left me feeling even more defenseless. "Jacket, pants, shirt, socks," Cinema demanded, holding out her hand.

I turned my face away as I took off each piece of borrowed clothing. I handed over Raffaele's sweatpants, his shirt and jacket, his white tube socks.

The cold traveled up from the concrete floor and through my bare feet. I hugged myself for warmth, waiting—for what? *What's coming next?* Surely they wouldn't give me a uniform, since I was a special case. It wouldn't make sense, since I'd be in prison so briefly.

"Your panties and bra, please," Lupa said. She was polite, even gentle, but it was still an order.

I stood naked in front of strangers for the second time that day. Completely disgraced, I hunched over, shielding my breasts with one arm. I had no dignity left. My eyes filled with tears. Cinema ran her fingers around the elastic of the period-stained red underwear I'd bought with Raffaele at Bubble, when I thought it'd be only a couple of days before I'd buy more with my mom.

Mom must be frantic. Is she still waiting at the train station? Wandering around Perugia looking for me? Has she called the police to help find me? Does she know I'm here?

"Squat," Lupa said.

I gave her a puzzled look.

She smiled encouragingly and bent her knees to show me. "You see?" she asked.

I squatted, and the women stared at me. Unlike at the *questura*, these guards were at least kind. They seemed almost like two distant aunts, looking at me with sympathy and speaking to me softly, knowing that what they were asking was excruciatingly humiliating.

Naked and crouching, cringing with shame, I held on to the knowledge that I would be released as soon as I could clear up the misunderstanding with the police. A few hours or maybe a day or two. No more than three—and for sure in a special holding cell, not in the real prison. I saw myself striding out of the gate in my hiking boots, book bag over my shoulder, Mom walking beside me, holding my hand.

"Now cough," Lupa said.

"What?" I asked, puzzled.

"Cough." She faked a cough. I imitated her.

"Good," Lupa said. "Here you go."

She handed me back my clothes, and I got dressed. But I was still shoeless. *"Che taglia di scarpe porti?"* she asked, pointing at my feet—"What size do you wear?"

"Porto una trenta-nove," I said softly—"I wear a thirty-nine."

"Go and look in the nuns' closet for something," Lupa told Cinema.

The female ward at Capanne had a chaplain and five nuns, who ministered to the inmates six days a week, filling in where the Italian government fell short—which included clothing us, since it turned out there weren't uniforms. The nuns kept a cabinet of donated apparel that they gave prisoners as needed—most of it worn out and poorly fitting.

Lupa pulled a lumpy, black plastic garbage bag out of a large bin and dropped it in front of me. It clanged against the floor. I dug down beneath the coarse, gray wool blanket folded on top and found a metal bowl and plate, a spoon and a fork, a plastic cup, a toothbrush and toothpaste, a plastic bag of gigantic feminine hygiene pads, a single roll of rough, brown toilet paper, and two sponges—one for scrubbing myself in the shower and the other for dishes. "Your provisions," Lupa said.

I choked at the back of my throat. I was holding a sack of the only things the government thought were essential to my life. I was in prison, and alone.

At that point, I gave myself as strong a pep talk as I could muster. *This is temporary—a stupid bureaucratic system that can't be bent. Like a roller-coaster ride that I've accidentally gotten on and can't get off until it's looped completely around. This is my own fault. I caused the confusion. Now I have to try to straighten it out.*

Tears streamed down my cheeks.

"Whoa! No, no. Be brave. You're okay," Lupa said.

Cinema came back carrying a worn pair of rust-colored cloth slippers, which I squeezed on over my socks.

"Those work okay," Lupa said, nodding approvingly. She held my upper arm, I gripped the garbage bag, and Cinema opened the door to the main hall. I walked out in a stranger's discarded house shoes.

Vice-Comandante Argirò, whom I'd met just before my strip search, was waiting. He was a thin man, probably in his fifties, with a large hooked nose that took up most of his droopy face and a hunchback that jutted out between his shoulder blades. He spoke his name loudly and slowly. "Ar-gi-rò," he'd said. "*Capi-sci l'i-taliano?*" Did I understand Italian? I nodded. "*Bene,*" he said, picking up speed. "*Sono vice-comandante. Capisci?*"

"*Sì,*" I said. Yes, I understood that he was a vice-commander and guessed that meant he was the second-highest person in charge of the prison.

When I'd first been brought inside from the squad car, I'd seen Raffaele through a barred glass window, locked in a hallway near the prison entrance. He was wearing his gray faux fur–lined jacket and was pacing back and forth, his head down. It was the first time since we'd been separated that I'd seen more than his feet. He didn't look at me. I'd wondered if he hated me.

Raffaele and I hadn't been together long, but I'd believed I knew him well. Now I felt I didn't know him at all.

I wondered why he was being kept here, what the police thought he knew, what the bureaucratic reasons were for his presence at the prison. I didn't know what was going on in Raffaele's head, but I imagined that he was as scared as I was. I couldn't imagine why he had betrayed me, but I wondered if he had been just as confused as I had been under interrogation, had lost faith in his own memories as well. Now I wonder if he realized then, unlike me, how serious our situation was.

I couldn't catch his eye before I was led away.

The next step was getting my mug shot taken. I was told to sit in a chair bolted to the wall and to look straight ahead into a big, black metal box, like the camera they use at the Washington Department of Motor Vehicles. In the second before they made the picture, it dawned on me that I wasn't supposed to smile for the camera. Later I was struck by how lost, how frazzled I appeared. My hair was wild. My skin was ghostly pale. My eyes were blank with exhaustion.

Argirò stared at the hickey on my neck, but said nothing. "Follow me, miss," he said finally.

Agente Lupa took my upper arm again and guided me forward. Argirò fit the large, gold key in his hand into the lock of a bulletproof glass door reinforced with a row of metal bars on either side. All the doors looked the same—and they were all impenetrable without a key. He went in ahead of us, holding the door open until we'd gone through, then closing and locking it behind us.

We walked through a series of dingy cream-colored hallways. Argirò unlocked each barred door as we went. Even in my daze, I noticed that none of the doors had knobs. The *vice-comandante* used his keys as handles.

I was inside the women's ward. Incredibly, as I went deeper and deeper into the cage, I didn't have the urge to escape.

Argirò led us up a narrow stairwell to *il primo piano*—"the second floor" (what we call the first floor is known in Europe as the ground floor)—and tapped the key he was holding against the barred door. On the other side, a female guard used her own gold key to unlock the door and usher us into the infirmary. With a patient's table in the middle of the floor, it was the first vaguely familiar-looking room I'd been in since walking into the prison. The doctor, an older man in a lab coat, his hair dyed dark, was sitting behind his desk. He looked down at the folder in front of him and up at me. "Name?" he asked.

"Amanda Knox. *K-n-o-x.*"

"Do you have allergies, illnesses, diseases?"

"No," I replied.

"Well, we'll need to do blood work anyway," he said. Just then I felt a sharp pinch from the back of my head. The nurse had snuck around me and plucked a hair from my scalp. I started to turn and glare at her, but instead asked the doctor, "Blood work? For what?"

"For diseases," he said. "Sign this. For the tests." He pushed a document and a pen in front of me, and I signed it. "How do you feel?"

"Worried," I said. "Worried and confused."

I shrank down in my seat.

"Confused?" he asked.

"I feel terrible about what happened at the police office. No one was listening to me," I said. Tears sprang to my eyes again.

"Hold up there, now," Argirò said.

"Wouldn't listen to you?" the doctor asked.

"I was hit on the head, twice," I said.

The doctor gestured to the nurse, who parted my hair and looked at my scalp.

"Not hard," I said. "It just startled me. And scared me."

"I've heard similar things about the police from other prisoners," the guard standing in the background said.

Their sympathy gave me the wrongheaded idea that the prison officials were distinct and distant from the police.

"Do you need anything to sleep?" the doctor asked.

I didn't know what he meant, because the idea of taking a sleeping pill was as foreign to me as being handcuffed. "No," I said. "I'm really tired already."

The doctor nodded to Argirò and the guards, and Lupa gently grabbed my upper arm, helping me to stand up. "Thank you," I said to the doctor.

I bristled at Lupa's touch, filled with resentment over being held on to like this. Did they think I might spontaneously do something horrible? But I forced myself to relax in her grip—I didn't want my anger and anxiety to be misread. From the start I tried to make it clear that their clasping my arm was unnecessary. The whole time I was at Capanne, I always spoke calmly and moved slowly, deliberately. When an *agente* grasped my arm, I imagined my arm shrinking until her fingers could encircle it without coming into contact with my skin. The assumption that I needed to be restrained like this made me furious. I didn't belong in a place where it was necessary to restrict people's movement by holding their biceps or handcuffing them because they might attack without warning. I didn't belong in prison.

Argirò led our procession up to *il secondo piano*—the third floor. "You're not to speak to anyone except the guards," he said. "No one but the guards." I guess he said it twice to make sure I got it. But who else could I have spoken to? There was no one else around. A tall, thin, red-haired female guard opened the next locked door. This hallway was lined with closed metal doors. I could hear the sounds of TVs and women's voices as we walked down the hall, but I saw no one until we came to the end. A pair of eyes peered out from the viewing window in the last door on the right.

The guard stepped forward and unlocked the last door on the left. Argirò went in first. He pointed at the TV, sitting on top of a gray metal box, opposite two beds. The TV was wrapped in brown paper and taped up, like a package waiting to be mailed. "Don't touch this," he said. "Don't you even try."

He must have been used to people who were much less compliant than I was. It wouldn't have occurred to me to disobey him. I felt oddly small, like Alice in Wonderland, when everything around her was so much bigger.

The bed looked as uninviting as you'd expect in a prison, with its yellow foam mattress on an orange metal frame pocked with black spots where the paint had chipped off. Two ugly burnt-orange metal cabinets were bolted into the wall—to hold clothes, I guessed.

"Take everything out of the garbage bag," the female guard said. "If you need anything call, '*Agente.*'"

"Am I allowed to make a phone call?" In movies, prisoners arc allowed one phone call.

Until that moment it hadn't occurred to me to ask. I needed to hear Mom's voice more than I'd ever needed anything in my whole life.

The guard looked at me like I'd asked for caviar and Prosecco.

I spread the thin blanket on the sheetless bed and lay down on the rough wool. I curled in on myself as the door clunked shut. "Stay here," the guard said, as if I had an option, and left.

With that, there was nothing left for me to do. I'd been at the beck and call of the police for five days and under their absolute control for nearly twenty-four hours. Being left alone was all I'd wanted during my interrogation. Now that I was, I was helpless and angry and terrified.

Now all I wanted was Mom. She had to be in Perugia by now, but I felt as far from her as I could ever be. I was sure she was freaking out about me, but there was nothing I could do. I wondered how she'd even find out where I was, what had happened to me. A horrible thought flitted through my mind: *What if she thinks I'm dead? That I've been killed, too? Like Meredith.*

I began to weep. Alone, I didn't even try to hold back.

My cell had its own bathroom and kitchen—two equal-size spaces together measuring about eleven feet by four feet and separated by a thick glass door. You had to go through the kitchen, just a long aluminum sink with an aluminum counter, to get to the bathroom, where the standard European fixtures—a sink, bidet, toilet, and shower—were lined up in a row.

Later, while I was sitting on the toilet, the redheaded guard came by and watched me through the peephole. So there was no privacy at all, then.

When I returned to the main part of the cell, someone passed a plastic plate through the barred door's single opening. Canned tuna, cut-up raw fennel, and rice smothered in tomato sauce. I had no appetite. I picked at the rice, which tasted comfortingly like the Uncle Ben's I was used to at home. I couldn't eat anything else.

I tucked myself back into the fetal position on top of the bed. A little while later, an *agente* walked by and closed the metal door over the bars. I thought, *I'm being sealed into a tomb.* Too claustrophobic and panicked to look in that direction, I rolled onto my other side to stare out the barred window into the dark.

Then I sobbed until I finally fell into a fitful sleep.

Chapter 13

November 7, 2007

I 'm not religious. I don't believe in miracles. I'm not sure what I think about God.

But the nun who visited me on my first full day in prison had an extraordinary effect on me. She was about eighty and wore a full habit, pale gray from head to toe.

She stuck both her hands through the bars of my cell and, grasping mine, told me, "*Dio sa tutto. Ti aiuterà a trovare la risposta.*"

Even though she uttered words I'd probably never have said in any language, I understood: "God knows everything. He will help you find the answer."

I had been alone in my cell, bent over a piece of white paper, a pen in hand, trying desperately to sort out the scraps of scrambled memories from the night of Meredith's murder, when the nun arrived. I had to remember exactly what I'd done the night of November 1. *Could the police be right? Did I have amnesia?* As the sister was leaving, she wished me *buona fortuna*—good luck. She smiled.

From the moment Meredith's body was discovered, I'd been searching the police's—and then the guards'—faces, silently pleading for reassurance that we were working together. But it was this nun with watery blue eyes, thinning gray eyebrows, and nearly translucent skin who gave me the strength to reconnect with myself.

Argirò had said this seclusion was to protect me from other prisoners—that it was standard procedure for people like me, people without a criminal record—but they were doing more than just keeping me separate. In forbidding me from watching TV or reading, in prohibiting me from contacting the people I loved and needed most, in not offering me a lawyer, and in leaving me alone with nothing but my own jumbled thoughts, they were maintaining my ignorance and must have been trying to control me, to push me to reveal why or how Meredith had died.

But I had nothing more to tell them. I was desolate. My scratchy wool blanket didn't stop the November chill from seeping bone deep. I lay on my bed crying, trying to soothe myself by softly singing the Beatles song "Let It Be," over and over.

I sat up when Agente Lupa came by my cell with another *agente* to check on me. "*Come stai?*" she asked.

I tried to answer, to say, "I'm okay," but I couldn't stop the surge of tears. Lupa asked her colleague to unlock the door and came inside. She squatted in front of me and took my cold hands in her large ones and rubbed them. "You have to stay strong," she said. "Everything will be figured out soon."

Then she hugged me like a mother does her distraught five-year-old. I buried my face in her shoulder and, in an explosion of emotion, bawled, as loudly as if I were screaming. I so desperately needed my mom that I took comfort from a stranger.

I ached to see my mother. A day had passed since she was supposed to have arrived, since I'd been out of contact. I could no longer fathom where she might be. I only knew that she must be trying to see me. She would get to me eventually. If only it had been sooner.

Six days ago I believed that I could, and should, cope with Meredith's murder by myself. But everything had broken down so quickly. I was sure that if I'd asked for Mom's help sooner, I wouldn't have felt so trapped and alone during my interrogation. I could have stopped it. If my mom, my lifeline, had been ready to jump to my defense on the other side of the door, I'd be staying with her now, not in prison by myself.

Lupa held me until my crying grew weak. "Do you need anything?" she asked.

"No," I whimpered. "Thank you."

When Lupa had gone, I returned to my scribbled memories about the evening of the murder. At the *questura*, when the police demanded I give them an hour-by-hour accounting of what Raffaele and I had done that night, I couldn't perfectly remember. We'd watched *Amélie*, eaten dinner, smoked pot, had sex, fallen asleep. But in what order? And what else? What had we talked about?

And then, right after the nun had left, detail after detail suddenly came back to me.

I read a chapter in *Harry Potter*.

We watched a movie.

We cooked dinner.

We smoked a joint.

Raffaele and I had sex.

And then I went to sleep.

What I'd said during my interrogation was wrong. I was never at the villa. I'd tried to believe what the police had said and had literally conjured that up. It wasn't real. That's not what happened. I hadn't witnessed anything terrible after all. I thought, *Oh, thank God!* I felt such a massive wave of relief.

I quickly wrote at the top of the page: "To the person who must know this."

Unlike my first *memoriale*, this one expressed less doubt and more certainty about where I'd been the night Meredith was killed. I rushed to get it down, so excited to finally be able to make sense of my memories for myself, and to be able to explain myself to the police. It read:

> Oh my God! I'm freaking out a bit now because I talked to a nun and I finally remember. It can't be a coincidence. I remember what I was doing with Raffaele at the time of the murder of my friend! We are both innocent! This is why: After dinner Raffaele began washing the dishes in the kitchen and I was giving him a back massage while he was doing it. It's something we do for one another when someone is cleaning dishes, because it makes cleaning better. I remember now that it was AFTER dinner that we smoked marijuana and while we smoked I began by saying that he shouldn't worry about the sink. He was upset because the sink was broken but it was new and I told him to not worry about it because it was only a little bad thing that had happened, and that little bad things are nothing to worry about. We began to talk more about what kind of people we were. We talked about how I'm more easygoing and less organized than he is, and how he is very organized because of the time he spent in Germany. It was during this conversation that Raffaele told me about his

past. How he had a horrible experience with drugs and alcohol. He told me that he drove his friends to a concert and that they were using cocaine, marijuana, he was drinking rum, and how, after the concert, when he was driving his passed-out friends home, how he had realized what a bad thing he had done and had decided to change. He told me about how in the past he dyed his hair yellow and another time when he was young had cut designs in his hair. He used to wear earrings. He did this because when he was young he played video games and watched Sailor Moon, a Japanese girl cartoon, and so he wasn't a popular kid at school. People made fun of him. I told him about how in high school I had been unpopular as well, because the people in my school thought I was a lesbian. We talked about his friends, how they hadn't changed from drug-using video game players, and how he was sad for them. We talked about his mother, how she had died and how he felt guilty because he had left her alone before she died. He told me that before she died she told him she wanted to die because she was alone and had nothing to live for. I told Raffaele that wasn't his fault that his mother was depressed and wanted to die. I told him he did the right thing by going to school. I told him that life is full of choices, and those choices aren't necessarily between good and bad. There are options between what is best and what is not, and all we have to do is do what we think is best. I told him that mistakes teach us to be better people, and so he shouldn't feel nervous about going to Milan to study, because he felt he needed to be nearer to his friends who hadn't changed and he felt needed him. But I told him he had to be true to himself. It was a very long conversation but it did happen and it must have happened at the time of Meredith's murder, so to clarify, this is what happened.

Around five in the evening Raffaele and I returned to his place to get comfortable. I checked my email on his computer for a while and then afterward I read a little Harry Potter to him in German.

We watched Amelie and afterward we kissed for a little while. I told him about how I really liked this movie and how my friends thought I was similar to Amelie because I'm a bit of a weirdo, in that I like random little things, like birds singing, and these little things make me happy. I don't remember if we had sex.

Raffaele made dinner and I watched him and we stayed together in the kitchen while dinner was cooking. After dinner Raffaele cleaned the dishes and this is when the pipes below came loose and flooded the kitchen floor with water. He was upset, but I told him we could clean it up tomorrow when I brought back a mop from my house. He put a few small towels over the water to soak up a little and then he threw them into the sink. I asked him what would make him feel better and he said he would like to smoke some hash.

I received a message from my boss about how I didn't have to come into work and I sent him a message back with the words: "*Ci vediamo. Buona serata.*"

While Raffaele rolled the joint I laid in bed quietly watching him. He asked me what I was thinking about and I told him I thought we were very different kinds of people. And so our conversation began, which I have already written about.

After our conversation I know we stayed in bed together for a long time. We had sex and then afterward we played our game of looking at each other and making faces. After this period of time we fell asleep and I didn't wake up until Friday morning.

This is what happened and I could swear by it. I'm sorry I didn't remember before and I'm sorry I said I could have been at the house when it happened. I said these things because I was confused and scared. I didn't lie when I said I thought the killer was Patrick. I was very stressed at the time and I really did think he was the murderer. But now I remember that I can't know who the murderer was because I didn't return back to the house.

I know the police will not be happy about this, but it's the truth and I don't know why my boyfriend told lies about me, but I think he is scared and doesn't remember well either. But this is what it is, this is what I remember.

I folded it up, gave it to the guard, and said, "I need this to go to the police."

I was a little girl again. I was doing what I'd done since I was seven years old, whenever I got into trouble with Mom. I'd sit with a *Lion King* notebook propped up against my knees, write out my explanation and apology, rip it out, fold it up, and then either hand it to Mom or, if I wasn't brave enough, put it somewhere I knew she'd immediately find it. When I was older I had a small, old-fashioned, beat-up wooden desk with a matching chair and a drawerful of pens. I felt so much more articulate writing than speaking. When I talk, my thoughts rush together, and I say things that don't always seem appropriate or make sense.

Writing brings order to my thoughts.

It always worked with my mom when I handed her my letter. She'd open it right away, while I stood by. She almost always cried when she read it. She'd hug me and say, "Thank you!" and assure me that everything was okay.

That's what I wanted to have happen now. Somehow the kindness from the nun and that embrace from Agente Lupa had encouraged me that it would.

I believed it was only a matter of time before the police understood that I was trying to help them and I would be released. The guard would unlock the cell. Without leading me by the arm, she'd escort me to an office where I could reclaim my hiking boots, my cell phone, my *life*. I'd walk out and into my mom's arms.

I thought I'd made it clear that I couldn't stand by what I'd said during my interrogation, that those words and my signature didn't count. We would have to talk again. This time they would have to listen and not shout.

I thought about what to do while I waited for my *memoriale* to get passed to the right readers and the paperwork to get filled in. Since I'd never been in a prison before—and I'd never be here again—I decided to record what I saw so I wouldn't forget.

I felt I had a duty to observe and collect information, just like a tourist who writes a travelogue or a war correspondent who witnesses devastation.

I inspected the gray-green paint on the walls, faded with age, and the splotches of white where the plaster was crumbling. A message had been left by a former occupant. Near the door, below eye level, in bright red lipstick, was an imprint of her puckered lips. Next to it, written in block letters, was a message: "*LIBERTÀ, SI ESCE, ESCO PRESTO*"—"FREEDOM, ONE LEAVES, I LEAVE SOON."

It was as though these words had been left for me. It was a message that added to my hope. I continued my inventory.

The barred window, about three feet by four feet, was thankfully large enough to let in light and allow me to look out onto the world I thought I'd left behind only temporarily. I saw a row of cone-shaped cypresses lined up on a hilltop. They reminded me of the trees Deanna and I saw two months ago, on the long, winding, and miscalculated hike from the Perugia train station—back when I'd been so sure of myself and so excited to see how my Italian adventure would unfold.

As I gathered this insider's information, I felt more like an observer than a participant. I found that being watched by a guard every time I peed or showered or just lay on my bed seemed less offensive when I looked at it with an impersonal eye. I saw the absurdity in it and documented it in my head.

But no matter how much I tried to distance myself from my physical surroundings, I was stuck with the anger and self-doubt that were festering inside me. I was furious for putting myself in this situation, panicked that I'd steered the investigation off course by delaying the police's search for the killer.

I thought back to the night of my interrogation—the police hovering over me, crowding in on me, pressing my cell phone in my face.

I imagined what I should have said: *"No! You're wrong!"*

That's what people believe they would have done in my place. They're certain they'd have held to the truth whatever the cost. They're certain they wouldn't have broken down and not known what the truth was anymore. That is what I would have imagined for myself: I would not have crumbled.

Chapter 14

November 8–9, 2007

Two nights passed with the metal door shut over my bars—an impenetrable shield that locked me inside alone. In the morning, when the guard opened it, nothing had changed. I was still in isolation.

I drank the coffee in the mornings, but barely touched the food that came around on a cart twice a day, delivered by a woman in a white apron and a net bonnet. It later turned out that she was an employed prisoner, but then I thought she must be from the outside. I kept trying to elicit a sympathetic look—to make a connection, however slight. But I got the same mechanical stare that I was getting used to seeing on almost everyone.

In the middle of my second full day as a prisoner, two *agenti* led me out of my cell, downstairs, outside, across the prison compound, and into the center building where I'd had my mug shot taken and my passport confiscated. There, in an empty office converted into a mini courtroom, seven people were waiting silently for me when I walked into the room, including two men, who stood as I entered.

Speaking in English, the taller, younger man, with spiky gray hair, said, "I'm Carlo Dalla Vedova. I'm from Rome." He gestured toward a heavier-set man with smooth white hair. "This is Luciano Ghirga, from Perugia." Each man was dressed in a crisp suit. "We're your lawyers. Your family hired us. The American embassy gave him our names. Please, sit in this chair. And don't say anything."

I was so grateful for my family's help. Finally I had allies, people to get me out of this unbearable situation.

Also in the room were three women. The one in black robes was Judge Claudia Matteini. Her secretary, seated next to her, announced, "Please stand."

In an emotionless monotone, the judge read, "You, Amanda Marie Knox, born 9 July 1987 in Seattle, Washington, U.S.A., are formally under investigation for the murder of Meredith Kercher. How do you respond? You have the right to remain silent."

I was stunned. My lower jaw plummeted. My legs trembled. I swung my face to the left to look at the only people I recognized in the room—Monica Napoleoni, the black-haired, taloned homicide chief; a male officer from my interrogation; and Pubblico Ministero Giuliano Mignini, the prosecutor, who I still thought was the mayor. Napoleoni was resting her chin on her hand glowering at me, studying my reaction. She seemed to be enjoying this.

Until the judge spoke, I had had no idea that I was being accused of murder.

I felt as though I'd been ambushed.

"Do you have anything to say for yourself?" the judge asked.

I turned to look at my lawyers. Carlo touched my hand and said, "Don't say anything now. We need to talk first."

There hadn't been enough time between their hiring and this preliminary hearing for Carlo and Luciano to meet with me. But more time might not have made a difference. It turned out that, mysteriously, Mignini had barred Raffaele's lawyers from seeing him before his hearing. Would the prosecutor have treated me the same? I think so. I can't be certain who ordered that I be put in isolation and not allowed to watch TV or to read, to cut me off from news from the outside world. But I believe that the police and prosecution purposely kept me uninformed so I would arrive at my first hearing totally unprepared to defend myself.

I do know this: if I'd met with my lawyers, I could have explained that I was innocent, that I knew nothing about the murder, that I imagined things during my interrogation that weren't true. The only thing my lawyers knew about me was that when I talked I got myself in trouble. I understand their impulse to keep me silent then, but in the end, my silence harmed me as much as anything I'd previously said.

When the judge asked if I had anything to say, I said no.

And with that one word I gave up my only chance to stand up for myself, my only chance to tell the truth.

I turned to my lawyers. "Please," I whispered urgently, "I have to explain."

"No, no, not right now," they said. "Don't say anything."

The whole hearing took less than ten minutes.

Just before I was taken back to my cell, Carlo said, "We'll come see you as soon as we can. And we're trying to work it out so that your mother can visit you."

Agenti led me back to the female prison. With each door that locked behind me I felt as if I were walking into a series of shrinking cages. I was trapped. Once I was in the smallest cage, my cell, I wailed.

It would be a long time before my Italian would be good enough to read Judge Matteini's nineteen-page report, which came out, and was leaked to the press, the next day. But my lawyers told me the gist of it. The judge said, "There were no doubts" that Patrick, Raffaele, and I were involved. Our motive, according to her, was that Raffaele and I wanted "to try a new sensation," while Patrick wanted to have sex with Meredith. When she refused, the three of us tried "to force her will," using Raffaele's pocketknife.

I couldn't believe anyone could think that of me.

The report continued: "It is possible to reconstruct what happened on the evening of November 1. Sollecito Raffaele and Knox Amanda spent the entire afternoon smoking hashish."

Judge Matteini claimed that I met Patrick at a "previously arranged" time and that Raffaele, "bored of the same old evening"—a phrase Raffaele had once posted online about himself—came along.

She went on to say that we hadn't called 112, the emergency number for the Carabinieri military police; that the Postal Police arrived at 12:35 P.M., and that our calls to 112 came afterward, at 12:51 P.M. and 12:54 P.M., suggesting that the police's appearance at the house took us by surprise and our calls were an attempt at orchestrating the appearance of our innocence. It wasn't until our trial that this accusation was proven to be erroneous.

The report said that in Raffaele's second statement, made on November 5, he changed his story. Instead of saying that we'd stayed at his apartment all night, as he'd done originally, he told police we'd left my apartment to go downtown at around 8:30 or

9 P.M., that I went to Le Chic and he returned to his apartment. He said that I'd convinced him to lie.

A bloody footprint allegedly compatible with Raffaele's Nikes was found at our villa, and the pocketknife he carried on his belt-loop was presumed to be compatible with the murder weapon.

The judge's report concluded that we "lost the appearance that [we] were persons informed about the facts and became suspects" when I confessed that Patrick had killed Meredith; that I wasn't sure whether or not Raffaele was there but that I woke up the next morning in his bed.

It was just the start of the many invented stories and giant leaps the prosecution would make to "prove" I was involved in the murder—and that my lawyers would have to try to knock down to prove my innocence.

———————

About an hour after I got back to my cell from the hearing, Agente Lupa came to the door. "Get your things," she said, smiling broadly. "I convinced the inspector to let you have a roommate so you're not by yourself. I'm moving you across the hall."

I was grateful for her effort. However, as I would end up sharing close quarters with a series of women I didn't know, often didn't like, and rarely felt I could trust, I couldn't help but remember the expression "killing with kindness." I sometimes mused to myself that the crazy roommates were another aspect of the prison intended by the prosecution to break me down.

My first cellmate, Gufa, was a woman in her late forties. She had decaying teeth and lank, greasy, graying hair. Her face and arms were covered with sores, which she picked at constantly.

Not knowing what they were, I was afraid to sit on the rim—there was no seat—of the toilet we shared, for fear she might be contagious.

The enormous glasses she wore made her look like an owl, and she sounded like one, too. When I walked in, she squawked at me in a dialect I could barely understand, telling me where to put my things, how to make my bed. She kept the room dark, because she slept off and on during the day. She collected garbage—food wrappers, pens without ink, used tissues—which she stored in her clothing locker, like a squirrel hiding nuts. Even though I never gave in to her, she was nosy, bossy, and demanding. She insisted on knowing about my case, wanted my lawyers to advise her on hers, and badgered me constantly to buy her snack food and other supplies and equipment for our cell from the order form that came around each week.

Still, she wasn't aggressive or spiteful, like other roommates I would eventually have. In her own weird way, Gufa tried to take care of me, in the same way a pet cat that drops a freshly dead rat at your feet thinks it's giving you a gift.

But what I wanted more than anything was for the guards to open the door and let me out. I wanted to see my mom. Until then, I just wanted to be left alone.

The day after my hearing, an *agente* did show up at my cell with a large gold key and unlocked the door. But it wasn't for the reason I'd hoped.

Grasping my arm, she led me down to a desk in the main hall on the ground floor. There, a curt, middle-aged guard pushed a piece of paper in front of me. "*Firmi qua, prego*," he said.

"Sign here? What is it?" I asked.

"It's the judge's paperwork," the male guard explained, his voice without inflection. "The confirmation of your arrest. It says the judge 'applies the cautionary measure of custody in prison for the duration of one year.'"

"One year!" I cried out.

I was floored. I had to sit down and put my head between my knees. That's when I learned how different Italian and U.S. laws can be. The law in Italy allows for suspects to be held without charge during an investigation for up to a year if a judge thinks they might flee, tamper with evidence, or commit a crime. In the United States, suspects have to be indicted to be kept in custody.

I felt I had only myself to blame. If I'd had the will to stick to the truth during my interrogation, I would never have been put in jail. My imprisonment was my fault, because I'd given in to the police's suggestions. I'd been weak, and I hated myself for it.

I sank to a new level of helplessness. Now I understood that no amount of explanation to the police or *memoriali* would clear things up. My fate was wholly dependent on the investigators. I wouldn't be released until the prosecution finished analyzing the evidence. Only then would they realize that I hadn't been at all involved in the murder.

Thank God I can count on my innocence to save me!

When they came to see me that afternoon, however, my lawyers weren't optimistic about a quick resolution. Likely they worried that I was stashed away in prison so the prosecutor, Giuliano Mignini, could steer the investigation in whatever direction he chose.

I veered between hope and despair. My life was in limbo. I couldn't do anything more to explain myself to the prosecution. I couldn't go back to school. I couldn't even ask Raffaele why he'd changed his story. I couldn't go home for Christmas. I felt as if the lights were going out. Sitting on my bed, hugging my knees, I thought, *I can't spend a whole year here. I'll die.*

Part Two

CAPANNE I

Chapter 15

November 10–13, 2007

The person I most needed to see was sitting alone at a wooden desk. Mom. Her straight hair fell limply around her face. As always, she was picking at her fingernails. I knew without looking that they were chewed raw, her reaction to anxiety. As soon as I walked into the small room, she stood, rushed toward me, and started crying, mirroring the relief I felt at having her in front of me.

Her face radiated concern and love, a look I'd seen when I came to her with a broken heart, or didn't do as well as I'd hoped on a test, or volunteered that I'd tried marijuana. Her empathy and advice always made me feel on safe ground. I didn't really get into trouble in high school, but I knew that if I did, she would support me through the situation. When I was at odds with myself, she'd reassure me that I was worthy of a happy life.

Now my no-questions-asked, I'll-come-help-you-wherever-you-are mother sat across from me in an empty room in Capanne Prison. This time she couldn't just make it all go away. She couldn't do anything but comfort me.

After being separated first by an ocean and two continents

and then by prison bureaucracy, walls, and bars, we hugged each other so tightly there was no space between us. Mom forced a smile, but tears rolled down her cheeks and into my hair as she nuzzled me. "Oh, my baby, my baby!" she whispered. "I love you so much. I've felt sick from needing to see you."

Holding me, she asked, "Are you all right? They say you told them you were there, that you were with Patrick. There are all these awful stories about you. Where are all these rumors coming from? Tell me everything."

"I'm so sorry, Mom. I'm so sorry," I moaned. "I didn't mean for any of this to happen."

I had so much to explain. After four days of being ordered around and ignored, I was finally in front of the one person who had always listened. But I worried that the overwhelming need I'd felt to tell the police what they wanted to hear wouldn't make sense to anyone who had never been pushed so far. How could I explain it to her when I didn't even understand it myself? More than anything, I needed my mother to believe me.

When we finally let go of each other, I pushed two chairs close together, fortunate that the *agente* observing us was less strict than some about letting visitors and inmates sit side by side. Like every room in the prison so far, this one was freezing. Mom held my hands in her lap and rubbed them. I knew the particular feeling of her hands—the only adult hands I'd ever come across as small as mine, but always warmer. She looked at me as though she were trying to absorb me into herself. Her face was strained from attempting to hold back tears. It tortured me to see Mom so upset and to know that I was the cause of it. But even troubled, her expressions—more familiar to me than my own face—were soothing.

I went through my interrogation with her step by step—the repeated questions, the yelling, the threats, the slaps. I explained to her how terrified I'd felt.

"I didn't come up with those things on my own," I said. "I told them I'd been with Raffaele all night at his apartment. But they demanded to know whom I'd left to meet, who Patrick was, if I had let him into the villa. They insisted I knew who the murderer was, that I'd be put in jail for thirty years if I didn't cooperate."

"Amanda," she said, her eyes wide, "I can't believe you had to go through that by yourself."

I told her that I had signed the witness statements out of confusion and exhaustion, that as soon as I had a few minutes by myself, I realized that what I'd said under pressure might be wrong. "I thought I could fix my mistake by explaining it in writing," I said. "Instead, they arrested me."

Mom listened, pulling me close. There was never a moment when it seemed that my words weren't reaching her.

When I finally finished, I got up the courage to ask her the question that had made me panic every time I'd thought about it. *What will I do if she says no or "I'm not sure"?*

I took a deep breath and exhaled. I was afraid to look her in the eye. "Do you believe me?"

I could see her surprise—then her sadness. "Of course I believe you! Oh, honey, how could I ever not?"

The immense burden I'd been carrying by myself lifted. I felt light-headed with relief. It was the first time since before my arrest that I'd talked to someone who knew I was innocent, who believed in me. I had longed to hear that for days—from anyone! Of course it came from the most important person in my life.

There was one more question I was burning to ask. "How did

you even know I was in prison? How did you know where to find me? They refused to let me pick up my phone when you called me. They wouldn't let me call you back."

"A friend of Chris's phoned the house after you were arrested," Mom said, tearing up again. "That's how he found out. My original flight to Rome was cancelled, and I'd just landed in Zurich when he called me. I just didn't believe it. As soon as we hung up, I ran to the bathroom and threw up. Then I had to call your father. It was the middle of the night in Seattle. I hated to leave such horrible news in a voice mail for him to wake up to, but I didn't have any choice. I had to let him know.

"I was frantic to get to you. And all I could do was wait in the airport for hours for a new flight to Rome. I didn't get to Perugia until about three A.M. By then you were already in jail. I was hysterical."

Since the hearing, I'd realized that she couldn't mamma-bear me out of prison. "Now I'll have to stay here until the prosecutor figures out there isn't any evidence against me—that I wasn't at the scene of Meredith's murder."

Mom squeezed my hands reassuringly. "I promise everything's going to be okay, Amanda. It's not your fault that the police scared you—you tried to fix things."

The *agente* opened the door. "It's time to go, Knox," she said. She pronounced my name the way all the prison officials did: "Kuh-nok-ks." It sounded as if they were trying to mimic the noise a hammer makes when dropped.

Mom held me tightly for half a minute more. We both were crying, but I felt so much better for having seen her. I didn't beg to stay with her longer, because I knew the answer would be no. Just asking would probably get me into trouble.

"I'll be back in a few days—as soon as they let me," Mom said. "Carlo and Luciano will come talk to you again, and your dad is flying over. This is all a big misunderstanding, and it will get fixed. We'll be here with you for as long as it takes. We'll get through this together. I love you so much."

I was led back to my cell, but my heart had lifted.

It is only recently that my mom confessed how wrenching our reunion was for her. Her words were barely coherent—just a rush of feeling: "Walking out of the prison without you that first time and many other times afterward was the hardest walking ever in my life—torture."

Three days later, my mom and dad came to *colloquio*—visiting hour—together. I remember thinking, when I walked in and saw them, *Things must really be bad if Dad's here.*

By then it wasn't as if I hadn't realized how much trouble I was in. But to see my father there added to the shock. I wasn't accustomed to having him involved in the nitty-gritty of my life.

My arrest had obviously shaken him. He seemed more tentative than I was used to. He didn't cry when I walked in, but he choked up and held me for a long time before letting go. He kept clearing his throat. His eyes were strained and red.

This was Dad's first trip outside the United States and the second time the three of us had sat down at a table together. I was a child again. My parents were making major decisions for me—and I knew the $4,800 I had left in the bank couldn't make up for the cost of two last-minute tickets from Seattle to Rome and the bills I was sure the lawyers were beginning to rack up.

My imprisonment didn't change the dynamic between Mom and Dad. They didn't suddenly seem like close friends. They didn't show affection for each other. They both focused on me.

But it made me swell with love for my parents to see that even though they were marked by their failed marriage, they were able to create a united front. They'd arranged this visit together. They were talking to Luciano and Carlo together. Inside an impersonal prison—stark white walls, harsh fluorescent lighting, a gray metal filing cabinet, and a cartoon drawing of Umbria—with a guard watching the clock, our time together didn't feel nearly as forced as our lunch at that café in Seattle. The three of us were sitting on the same side of the table, our chairs squeezed as close together as possible. I was in the middle, with each parent gripping one of my hands.

Capanne made eight hours available for visitors each month—on Tuesdays and Saturdays—but the prison allowed each prisoner only six visits. This infuriated my parents, who wanted to be there each time the prison was open to outsiders. It made me crazy, too. Eventually Carlo and Luciano were able to arrange eight *colloqui* a month, and sometimes nine, by pleading with the prison authorities that my family had to come so far to see me. Even with the bumped-up hours, the amount of time I was able to spend with the people I loved was such a tiny fraction of the thousands of hours I was locked up, trapped among strangers.

What my family ultimately managed to do for me, while living nearly six thousand miles away, was incredible. I'm sure I had more support than most of the inmates, including the ones who grew up down the street from Capanne. There was hardly a time that someone—Mom, Dad, or Chris—wasn't there, unless they'd arranged for an aunt, uncle, or friend to sub in.

There was nothing pleasant in it for them. They rented a tiny apartment in the countryside, about ten miles away from Capanne, left their spouses and my sisters, put their lives on hold, and took turns staying in Italy for weeks at a time. They didn't speak the language or know another soul. They came to Perugia for one reason: to see me for one hour, two times a week.

Without them, I think I would have had a complete breakdown. I would not have been able to survive my imprisonment.

Before my parents left together that first time, Mom grasped my hands again, leaned toward me, and, tears brimming, said urgently, "Amanda, I'd do anything to take your place. Your job now is to take care of yourself. I'm worried for you being here."

Her words underscored what we all knew: that while my parents had my back, they couldn't take care of me from day to day. I had to navigate prison alone. For other prisoners, the key to survival was to find someone to bond with, and that person would protect you and guide you through. But there was no one like me, no one I could confide in, no one whom I could trust to take me under her wing.

Chapter 16

November 9–14, 2007

The best part of my day was the few seconds between waking and remembering. During that moment, with my eyes not yet open, I was in my cozy lemon yellow room at Mom's house in Seattle. I was happy.

Then I'd remember that I was locked in a cold cell where the radiators were turned on for only a few hours a day. And panic would overtake me. *How can I be in prison? How can I be accused of something so horrible? It seems impossible. Yet here I am.*

Getting up, I'd look out the barred window and envy the rabbits hopping across the empty, dank fields. I wasn't even exactly sure where Capanne was—all I knew was that it was somewhere between Perugia and Rome.

Some days I felt as if I were in limbo, because I wasn't able to connect to the real world. I was adrift. Mom and Dad were my anchors, and I measured time by their visits. *Two days until they're here. Four days until they can come back.*

In spite of all that had happened, I believed that the police, the prosecutor, a judge—some official—would look at the facts and realize how wrong they'd been. They'd be jolted by the obvious: that I was incapable of murder. Surely someone would see that there was no evidence. My belief that my imprisonment was temporary was all that kept me from being overwhelmed. I guess my faith in eventual justice is what psychologists call a coping mechanism.

In the days after Meredith's death I'd insisted on staying in Perugia. Back then, going home meant defeat. But my wants flipped with my arrest. Now the only thing that mattered was to reclaim my life in Seattle. I considered what I would do once my ordeal was over—how I'd rebuild myself, whether I'd live with Mom or find a place of my own, whether I'd go back to school or get a job, how much I wanted to reunite with the people I loved.

I was determined not to settle in at Capanne. I saw that as a victory for the officials who thought I was guilty. I told myself I'd leave no trace of having been there; I'd carry out only what I'd carried in—a lesson my family had taught me when we went camping. In my mind, I *was* camping. *This is no more permanent than a week in the mountains*, I told myself. My only possessions were the few impersonal supplies that came in the garbage bag I was handed the first night and a few utilitarian items from the nuns' closet—sheets and a stiff bath towel. I was determined to make do. The idea of getting comfortable was terrifying.

Mom begged me to tell her what I needed. "To leave this place," I said. But knowing I couldn't, I asked for a couple of pairs of underwear and a few T-shirts.

A guard gave me an order form for groceries and other basics—ranging from salt to sewing needles—and a *libretto*, an eight-and-a-half-by-eleven-inch piece of paper folded in half with a handwritten spreadsheet inside to track what I spent. I had two hundred euros—about three hundred dollars—in my prison account from the purse/book bag they'd impounded upon my arrival. The order form was divided into three columns for the name of the item, the code number, and the quantity. Gufa badgered me to buy her a camp stove and a coffeemaker, but I refused to order so much as a carton of milk. I'd be gone before it reached its expiration date.

Getting me out of jail was the first priority whenever I talked to Carlo and Luciano. Their take was that when the media frenzy died down in a couple of weeks, a judge would probably put me under house arrest, either with my family or in a religious community. Then, when the prosecution saw they had no evidence against me, they would let me go.

As the days crept by, though, I renegotiated my deal with myself. *Amanda, you're going to need a few things. Buying won't mean you're staying.*

I filled in the columns for a toothbrush, toothpaste, and a hairbrush.

A few days later a short, thin young woman dressed as an adolescent—in jeans, a sweatshirt, and Miss Piggy sneakers—brought me my order, passing it through the meal slot in the bars. One of the items completely baffled me. "No, no," I said, realizing what it was. "I want it for the hair."

"Oh," she said. She laughed good-naturedly and showed the guard my mistake. I was mortified, as I always was, when my ignorance tripped me up.

In filling out the order form, I'd requested a men's shaving brush—a *spazzola da barba*—instead of a *spazzola per i capelli.*

"Don't worry, I'll see if I can exchange it," she offered.

"Thank you," I said. "You're so kind to prisoners."

She laughed heartily this time and caught Lupa's eye. "Fanta *is* a prisoner," Lupa explained to me. "All the workers you see are prisoners."

It wasn't just the language that threw me. Almost every aspect of life at Capanne was foreign. (The garbage bag I'd been given upon my arrival hadn't come with a prison "user's manual.") Gufa quickly nicknamed me "Bimba"—"little girl." She said it in a playful way, but at the same time it underscored how clueless I was.

I was at the mercy of my jailers. I had no idea what to anticipate or how to act. What should my relationship with other prisoners, guards, prison officials, be? How open could I—or should I—be and with whom? As naïve as I now realize this was, when guards and prison officials, psychologists and doctors, asked me about myself, I didn't know if I was allowed to keep my thoughts private or if I always had to tell them exactly what was on my mind.

I wondered about the basic rhythm of things. How was I supposed to wash my clothes? How did I perform the essential routines of daily life? And whom should I ask? I found out that in order to get an appointment with the prison *comandante*, to buy anything that wasn't on the grocery list, to switch out clothing or books in the storage room, to get a prison job, to pass your belongings on to another prisoner, to change cells—for nearly everything—you had to fill out a *domandina*, to ask permission.

No one explained to me how anything worked unless I made a mistake. When my family brought me a puffy ski jacket, I found out that padded material was off-limits, apparently because drugs could be hidden inside. Many items were on the "No" list for this reason. Among them: comforters, soft cheese, homemade cookies, and some types of buttons. Even nutmeg was forbidden. Apparently, when eaten in large quantities or smoked, it can make people drunk or high. Gloves were allowed only if the fingers had been cut off. When I got mail, a guard would bring the envelope to my door and open it in front of me. She always tore the stamps off my letters—drugs could be glued on the back—and gave the letters to me page by excruciating page. If I wanted the envelope, it had to be checked first for poison, razors, and, of course, drugs.

During the first month, I found out that most *agenti* kept an emotional distance from prisoners. Many would ask you about yourself but would never tell you their name or anything about their lives outside prison. One day, when a guard called Rossa was walking me upstairs from a visit with the doctor, I asked her, "Are you having a good day?"

"You need to stop kidding yourself and acting like we're friends," she snapped. "I'm an *agente*, and you're a prisoner. You need to behave like one. I'm doing you a favor by warning you."

I felt my face go red, humiliated by the reality of my situation.

One of the few things that didn't upset me was Capanne's clockwork consistency—coffee, tea, or milk at 7:30 A.M., lunch at 11:45 A.M., dinner at 5:45 P.M. The routine helped the days blur together and the waiting go faster.

But time stretches in prison. I was awake at least sixteen tedious, empty hours a day—with few options for filling them. I tried to block out my claustrophobia with reading, writing, and sit-ups. Lupa had rummaged through the prison book closet to bring me *Harry Potter and the Goblet of Fire* in Italian, along with an Italian grammar book, and a dictionary. I still cared about learning Italian, even then, and I spent hours looking up definitions and diagramming each sentence into subject and predicate. Anything that made me feel purposeful gave me emotional comfort, and it was psychologically essential for me to find a silver lining in my imprisonment. Later, learning Italian became more about self-defense and survival: I had to speak Italian if I wanted to communicate and, ultimately, defend myself.

Early on, I started keeping a journal, which I titled *"Il mio diario del prigione"*—"My Prison Diary"—on the cover:

> *My friend was murdered. My roommate, my friend. She was beautiful, smart, fun, and caring and she was murdered. Everyone I know is devastated for her, but we are also all at odds. We are angry. We want justice. But against who? We all want to know, but we all don't . . .*
>
> *Now there's the sound of women wailing through bars and the sounds of wheels of the medicine carts rolling down the hard floors of the echoing halls.*
>
> *November 2007*

But I spent most of my time sitting on my bed wondering what was happening beyond the sixty-foot-high walls topped with coiled razor wire. What were my parents and family and friends doing and thinking? What was happening with the investigation? How long would it take to examine the forensic evidence that would clear me?

Underneath every thought there was a bigger, louder one looping through my head. *How could I have been so weak when I was interrogated? How did I lose my grip on the truth? Why didn't I stand up to the police?* I'd failed myself, Meredith, Patrick, Raffaele.

Just about the only relief from the excruciating boredom and relentless self-criticism was *passeggio*, the hour a day I got to leave my cell and exercise outdoors. Because I was still separated from the prison population, and Gufa didn't take full advantage of her exercise periods, I didn't realize that other prisoners could go outside twice a day for two hours at a time. Even if they didn't want to work out, it was an opportunity to socialize that I didn't have.

I was being treated differently from the other prisoners. While I exercised, I could see other inmates through the barred glass doors, chatting and moving around freely inside without an *agente*. A guard watched me at all times. There was no conversation.

During that hour I also noticed toddlers being led through the halls by a nun. The sight of them delighted and perplexed me. Who were they? Where had they come from? I thought they might be orphans, but when I asked an *agente* about them, I discovered that the female prison had a separate *nido*—literally "nest," or nursery ward—where women lived with their children until the children turned three. It made sense not to separate mother and child, but why did they live in prison? Couldn't these women be held under house arrest, or in religious communities, so the little kids wouldn't have to be behind bars?

For the first few days, the toddlers I saw were shy. They either scampered off or stood watching me wave to them, their hands in their mouths, expressionless. I didn't know where the nun was taking them, but I was thrilled that their passage coincided with my outside time. After a couple of days the children in the mini parade would stop and watch me through the bars. I tried everything to make them laugh—I danced, sang, played peek-a-boo, and we chased each other on opposite sides of the window, with them on the inside, me on the outside. Some days, the nun had to cajole them into leaving.

My *passeggio* was in a small courtyard outside the chapel—

really just a wide path surrounding a muddy, round patch of a garden with a crudely done abstract sculpture in the center. I could never decide if the matte-gray metal blob was supposed to be two wings or splashing waves rising out of the ground and tipping toward each other. But I was sure of one thing: it was ugly.

Prison was not the place to find inspiration.

I exercised as much to stay warm as to stay in shape. Breaking a sweat cleared my mind and tamped down my anxiety. After I exhausted myself, I'd walk in the tight, hypnotic circles available to me, singing or repeating the mantra *It's going to be okay. Just hold on. It's going to be okay.* Or I'd pace and cry, remembering how scared I was during my interrogation, remembering fragments of my time with Meredith and trying to process her death. And I thought about my family; I hated that I'd put this burden on my parents.

No matter what mood I was in, I'd stoop to pick up earthworms that were washed onto the pavement when it rained and lay them gently back in the dirt.

The person inside the prison who came closest to taking care of me was Don Saulo Scarabattoli, the Catholic chaplain for Capanne's women's ward. A few days after I moved in, he appeared at my cell door, introduced himself, and asked, "Would you like to come talk with me in my office?" He was smiling and grasping my hands through the bars.

"I'm not religious," I said. "I wouldn't really have much to say."

"That's not a problem. You're always welcome," he answered.

A short man with a large bald spot and a little gray hair, Don Saulo was in his early seventies. His wire-frame glasses and the white stubble on his jaw made him appear gentle. But I was put off by the small cross pinned on his navy blue sweater and the Virgin Mother medallion around his neck. I was not interested in being converted.

I wasn't baptized as a baby. Growing up, I never went to Sunday school, never said grace before meals, never prayed before bed. I stereotyped religion as a backward institution that offered false comfort and prevented people from coming to their own conclusions. Early in my freshman year at my Jesuit high school, I figured out that by showing up late on Fridays I could ditch Mass without repercussions, and I gave short shrift to my required religion classes. Once, when we were assigned a paper on how our belief in God had influenced our life, I curtly wrote that it hadn't affected mine because I didn't believe in God. My tone made it obvious that I thought the assignment was inane, and earned me my only C in high school, which frustrated me even more. Who was my teacher to grade me on my personal beliefs?

I assumed that people who dedicated their lives to religion were trained to be nice—that it was part of their professional code of conduct and not always authentic. I had a cautious reaction to Don Saulo even after he said that the priesthood bound him to keep in confidence whatever I told him. I didn't trust him. My lawyers had warned me that whoever was asking me questions was likely to be a police informant.

But I quickly came to believe that Don Saulo's decency and compassion were genuine. This made our first few conversations painful for me. Each time, I'd ask him, "Do you believe I'm innocent?"

"I believe you are sincere," he would answer. Tears rolled down his cheeks and mine.

Sincere is not the same as innocent. But I soon stopped needing that affirmation from him. I could tell that he was an intelligent, caring person interested in me as another human being. He didn't push to know about my case or even to get me to talk.

Of course, Don Saulo did offer me many opportunities to join the Catholic Church, and God came up in all our conversations. Although I still didn't believe in an omnipotent being or revere any faith as inarguable Truth, I slowly moved away from the rebellious stance I'd taken in high school. Instead of thinking that religion inhibited individuality, I came to see it as the collective wisdom of countless generations. I respected it as a way to examine fundamental questions. What does it mean to be human? What defines a good life? Why do we exist? Religion was Don Saulo's language, and it gave me a way to talk with him about my feelings, insecurities, and ideas.

I tried not to discuss my case with anyone but Mom, Dad, and my lawyers. But I confided in Don Saulo about my interrogation, how guilty I felt about naming Patrick, and how confused I was by what had happened to me. "If you did not knowingly wrong someone, Amanda," he said, "you did not sin."

And Don Saulo did look out for me. Every Tuesday he screened a movie in the women's ward under the guise of "rehabilitation." To my amazement, he convinced prison officials to let me attend. The theater was a large room, empty except for rows of plastic chairs and a piano, which we weren't allowed to touch, and a retractable movie screen.

Movies, like everything else, brought out Don Saulo's emotional side. I can't remember the lights once coming back on at the end of a movie when his cheeks weren't wet and his voice didn't quaver, whether we'd just watched *The Passion of the Christ*, *Bruce Almighty*, or *The Princess and the Frog*.

And he burst out laughing as easily as he cried. One day, when we were talking in his office, I shifted in my hard chair, clearly uncomfortable. "Are you all right?" he asked.

"My *culo* hurts," I answered.

He chuckled. "You mean your *sedere* hurts," he corrected me, smiling.

Culo—"ass"—is a word I'd picked up from Gufa. No one would use such a vulgar term in front of a priest unless she were trying to offend him. I was embarrassed, but Don Saulo was amused.

My visits with him were optional. I saw Don Saulo for half an hour a few times a week because he was the only person I met at Capanne who liked sharing and debating ideas. Most conversations I had were with people I didn't like—my cellmate, some doctors, Vice-Comandante Argirò, the cops who came to Capanne to confiscate more and more of my things. I had no choice but to speak with them. Not being able to choose where I went and whom I saw made me anxious. It seemed as if everyone around me was trying to chisel their way into my head. Even the letters I wrote had to be turned over to the guard in an unsealed envelope—to be photocopied for the police, I later discovered. I felt I had to protect myself from invasion.

Each morning, after the other prisoners had filed outside for *passeggio*, an *agente* would escort me to the infirmary for the first of my required twice-daily visits. All prisoners are under some sort of observation, but someone—probably the prosecutor— had ordered that the doctors question me regularly for my first six months at Capanne in the hope that I'd say something incriminating.

It would never have occurred to me to take anything to improve my mood or help me sleep, but almost every doctor recommended antidepressants and sedatives. The psychiatrist seemed particularly determined to get me to succumb to drugs. I countered with an emphatic "No!" each time. I wasn't about to give the prison officials *more* control over me than they already had.

Doctor-patient confidentiality didn't exist in prison. A guard was ever-present, standing right behind me. This bothered me so much that, as time went on, I skipped a needed pelvic exam and didn't seek help when I got hives or when my hair started falling out. Whatever happened in the infirmary was recycled as gossip that traveled from official to official and, sometimes, back to me.

How each visit went depended on the doctor, and I was grateful for any gesture that wasn't aggressive or disdainful. A female physician liked to talk to me about her trouble with men. And one day, when I was being seen by an older male doctor, he asked me, "What's your favorite animal?"

"It's a lion," I said. "Like *The Lion King—Il Re Leone*."

The next time I saw him he handed me a picture of a lion he'd ripped out from an animal calendar. I drew him a colorful picture in return, which he taped to the infirmary wall. Later, when he found out that I liked the Beatles, one of us would hum a few bars from various songs to see if the other could name the tune.

But sometimes what I thought was a kind overture would take an ugly turn. I was required to meet with Vice-Comandante Argirò every night at 8 P.M. in his office—the last order before lights out at 9 P.M. I thought he wanted to help me and to understand what had happened at the *questura*, but almost immediately I saw that he didn't care. When I ran into him in the hallway he'd hover over me, his face inches from mine, staring, sneering. "It's a shame you're here," he'd say, "because you are such a pretty girl," and "Be careful what you eat—you have a nice, hourglass figure, and you don't want to ruin it like the other people here."

He also liked to ask me about sex.

The first time he asked me if I was good at sex, I was sure I'd misheard him.

I looked at him incredulously and said, "What?!"

He just smiled and said, "Come on, just answer the question. You know, don't you?"

Every conversation came around to sex. He'd say, "I hear you like to have sex. How do you like to have sex? What positions do you like most? Would you have sex with me? No? I'm too old for you?"

His lewd comments took me back to the pickup lines used by Italian students when I'd relax on the Duomo steps in Perugia. I wondered if I should just chalk up his lack of professionalism to a cultural difference. Sitting across the desk from him, I thought it must be acceptable for Italian men to banter like this while they were on the clock, in uniform, talking to a subordinate—a prisoner.

He had me meet with him privately and often showed up during my medical visits, but I had always been so sheltered, I didn't think of what he did as sexual harassment—I guess because he never touched or threatened me.

At first when he brought up sex I pretended I didn't understand. "I'm sorry—*Mi dispiace*," I'd say, shaking my head. But every night after dinner, I felt a knot in the pit of my stomach. I had no choice but to meet with him. After about a week of this behavior, I told my parents what Argirò was saying. My dad said, "Amanda, he shouldn't be doing that! You've got to tell someone!"

Knowing that Dad thought this was wrong validated my own

thoughts. But Argirò was the boss—what could I do? Whom could I tell? Who'd take my word over his?

Silently, I rehearsed what I would say to him: "These conversations repulse me." But when we were face-to-face, I balked, settling on something more diplomatic—"Your questions make me uncomfortable," I said.

"Why?" he asked.

I thought, *Because you're an old perv.* Instead I said, "I'm not ashamed of my sexuality, but it's my own business, and I don't like to talk about it."

It didn't do a bit of good. He ended that night's meeting telling me that my hair looked nice. It was in a ponytail. He tried to hug me before I left. I backed away.

I still wasn't sure this was something I should bother Luciano and Carlo with. But when it continued for a few more days, I did.

Luciano looked revolted, and Carlo urged me, "Anytime Argirò calls you alone into an office, tell him you don't want to speak with him. He could be talking about sex because Meredith was supposedly the victim of a sexual crime and he wants to see what you'll say. It could be a trap."

But I was so lacking in confidence I couldn't imagine it would be okay to resist Argirò directly. I reminded myself that the pressure I felt during these sessions wasn't anything close to the pressure I'd been put under during my interrogation. Argirò usually sat back and smoked a cigarette, and I knew that I could just wait out his questions. Eventually he'd send me back to my cell. I didn't tell him off because I'm not a confrontational person. When something bothers me I try to ignore it and get over it or address it in a roundabout way. That's why I wrote all those apology letters to my mom when I was young, instead of approaching her outright.

One night, Argirò asked me if I dreamed about sex, if I fantasized about it.

Finally I got up my courage. I took a deep breath. "For the last time," I said, my voice pitched, "No! Why are you constantly asking me about sex?"

Argirò stared and shrugged, like it was no big deal—that it was my fault for not drawing the line in the first place.

Chapter 17

November 15–16, 2007

Vice-Comandante Argirò broke the news. Instead of his usual greeting—a lecherous smile and a kiss on both cheeks—he stayed seated behind his desk. His cigarette was trailing smoke. His face was somber. Something was wrong.

He pushed a printout of an Italian news article toward me. It took me a minute to translate the headline: "Murder Weapon Found—With DNA of Victim and Arrested Suspect Knox." Beneath was a fuzzy photograph of a kitchen knife and the words "A knife has been found in Sollecito's apartment with Knox's DNA on the handle and the victim's DNA on the blade. Investigators believe it to be the murder weapon."

That doesn't make sense. I must have read it wrong.

I made myself start over, slowly rereading the story, checking each word as I went. By the end I knew language wasn't the barrier.

Argirò glared at me cruelly.

"Do you have anything to say?" he asked.

"It's impossible!" I blurted. "I didn't kill Meredith! I'm innocent! I don't care what the article says! It's wrong!"

"It's proof," Argirò said, smirking. "*Your* fingerprints. *Her* DNA."

"I don't know anything about a knife," I said. "You can't prove that I'm guilty when I'm innocent."

The short conversation ended in a stalemate. I glowered at him.

"Why don't you go back to your cell and think about what you want to say," Argirò said.

I didn't have any words for my anger—or fear. They were roiling inside me as the *agente* led me away.

Maybe I shouldn't have been shocked by the accusation about the knife. I'd been in jail for nine days, and I'd already been billed as a murderer.

My lawyers had to keep coming to Capanne to relay the ever-changing story of my supposed involvement and ask me if there was truth to the reports.

Investigators were claiming that I'd been responsible for holding Meredith down while either Patrick or Raffaele cut her throat, that I'd pressed so hard on Meredith's face during the attack I'd left an imprint of my fingers on her chin. The police said that because the bruises were small, they'd come from a woman's fingers, even though that's not how it works. "It isn't like a fingerprint," Carlo explained. "You can't tell the size of the hand by the size of the bruise. It depends on the circumstances and the pressure."

This was another example of the prosecution misinterpreting evidence so it would put me at the murder scene and discounting the things that didn't fit into their explanation. They had done the same thing a few days before, when they circulated the idea that only a woman would have covered Meredith's ravaged body with a blanket. A few years later I learned that this is something first-time killers also often do. The detectives didn't mention how improbable it is for a woman to commit a violent crime, especially against another woman. Nor did they acknowledge that I didn't fit the profile of a violent woman. I'd never been in a gang; I had no history of violence.

The untruths kept coming—seemingly leaked from the prosecutor's office.

In mid-November the press announced that the striped sweater I'd worn the night of the murder was missing, implying I'd gotten rid of it to hide bloodstains. In truth I'd left it on top of my bed when I came home to change on the morning of November 2. The investigators found it in January 2008—in the same spot where I'd taken it off. It was captured in photos taken of my room, which my lawyers saw among the official court documents deposited as the investigation progressed. The prosecution quietly dropped the "missing sweater" as an element in the investigation without correcting the information publicly. Convinced that arguing the case in the media would dilute our credibility in the courtroom, Carlo and Luciano let the original story stand.

Things that never happened were reported as fact.

The tabloids said I'd met a nonexistent Argentinian boyfriend in a Laundromat to wash my bloodstained clothes. *False.*

The Italian news channel reported that cameras, mounted on the parking garage across the street from the villa, captured a girl dressed in a colored skirt or blouse, presumed to be me, emerging from the garage at 8:43 P.M. the night of the murder. *False.*

The police leaked this to the local press, and it rippled out from there. If true, it would have contradicted my alibi: I hadn't left Raffaele's apartment that night. The local headlines in those days often read "Amanda *Smentita*"—"Amanda Found in a Lie." It bolstered the prosecution's characterization of me as a depraved, deceitful person capable of murder.

Later, investigators decided the video image wasn't sharp enough to decipher, that it would be too easy for the defense to knock down. But the damage had already been done.

The press reported police claims that Raffaele and I had destroyed the hard drives on four computers—his, mine, Filomena's, and Meredith's. *False.*

Later, when a computer expert examined the computers, he discovered that the police had fried the hard drives. Whether it was on purpose or out of extraordinary incompetence, I never learned. But it's hard to see how they could inadvertently have wiped out four computers, one after the other. My computer wouldn't have given me an alibi. All investigators would have found was evidence of Meredith's and my friendship—pictures from the Eurochocolate festival and of our hanging out at home.

Journalists reported that the police had confiscated "incriminating" receipts for bleach, supposedly from the morning of November 2. *False.*

The receipts were meant to show that Raffaele and I had bought bleach—what Americans call household cleaner—and spent the night of the murder cleaning up the crime scene.

Four of the receipts were dated months before I arrived in Perugia, and bleach wasn't among the items purchased. The last one was from November 4, two days after Meredith's body was found. And it wasn't for bleach. It was for pizza. But no press corrected the story or reported the truth.

There seemed to be an endless chain of headlines, like new pieces of candy to wave in front of people. New evidence!

Amanda said this! As soon as the police fed them a new tidbit of unfounded news, the earlier headline would be replaced. The media seemed less interested in investigating the claims than in just hanging them out there. And the tabloid sensationalism of one country was recycled to become legitimate news in another.

Still, none of the investigators' claims was as unfathomable to me, as damning, as the reports about the knife.

When I read the article in Argirò's office, it seemed as fake as a grocery store tabloid claiming "Martian Baby Born in 7-Eleven Has Three Heads."

Sitting in his cold office, staring at the printout, I could think of only two ways the knife news had come to be. Choice one was that the website had fabricated it. As dishonest and unprofessional as the media had been, I was pretty sure they wouldn't go this far. Choice two was that the investigators had made a mistake.

I went over what I knew, step by step.

A knife from Raffaele's kitchen with DNA from both Meredith and me wasn't possible. In the week I'd known him, I'd used Raffaele's chef's knives to cook with, but we had never taken them out of his kitchen.

Meredith had never been to his apartment.

But I could present my argument only to myself and to Argirò.

I could tell that Argirò didn't believe me. I knew the knife could not be the one used by Meredith's killer. My heart felt as if it were being squeezed.

Back in my cell, I was quiet and withdrawn, spending the rest of the night venting to my prison diary. (That would end up being a mistake.) I told Gufa I was too tired to talk. I couldn't sleep.

Luciano and Carlo arrived the next morning. "Are the police

really claiming they found a knife in Raffaele's apartment with my DNA on the handle and Meredith's DNA on the blade?" I asked, desperate for them to say no.

"The police are saying that the knife is the murder weapon," Carlo said. "Their forensic experts believe that it was capable of inflicting each wound on Meredith's body. They've given up the idea that it was Raffaele's pocketknife. Amanda, they're saying you're the one who stabbed Meredith. Is there something you need to tell us?"

Both men looked at me intently, gauging my reaction.

I couldn't believe what they were asking me. "No! It's impossible!" I shrieked, my body starting to shake. "The police have made a mistake. I never left Raffaele's that night, I never took a knife from his apartment, and Meredith never visited me there. I didn't have any reason to be angry with Meredith. And even if we'd had a fight I would have talked to her, not killed her!"

"We believe you, Amanda," Carlo said immediately. "Don't worry."

Investigators apparently had confiscated the knife—a chef's knife with a black plastic handle and a six-and-a-half-inch blade—when they searched Raffaele's apartment after our arrest. It was the only knife they considered out of every location they'd impounded, the top knife in a stack of other knives in a drawer that housed the carrot peeler and the salad tongs. I'd probably used it to slice tomatoes when Raffaele and I made dinner the night Meredith was killed.

The officer who confiscated the knife claimed that he'd been drawn to it by "investigative intuition." It had struck him as suspiciously clean, as though we'd scrubbed it. When he chose it, he didn't even know the dimensions of Meredith's stab wounds.

The knife was a game changer for my lawyers, who now feared that the prosecution was mishandling evidence and building an unsubstantiated case against me. Carlo and Luciano went from saying that the lack of evidence would prove my innocence to warning me that the prosecution was out to get me, and steeling me for a fight. "There's no counting on them anymore," Carlo said. "We're up against a witch hunt. But it's going to be okay."

They were confident that once our forensic consultants could show how wrong the prosecution was, we would ultimately win. But I also think their promises were meant to keep me from spinning into crisis, especially since I only saw them once a week.

I was choked with fear. The knife was my first inkling that the investigation was not going as I'd expected. I didn't accept the possibility that the police were biased against me. I believed that the prosecution would eventually figure out that it wasn't the murder weapon and that I wasn't the murderer. In retrospect I understand that the police were determined to make the evidence fit their theory of the crime, rather than the other way around, and that theory hinged on my involvement. But something in me refused to see this then.

Soon after the knife news came out, the police came to the prison to confiscate my purse/book bag. I was called down to *la piano terra*—"the ground floor"—to witness the seizure from storage and once again sign a document. They took what was left of the bag's contents after the interrogation—my textbooks and notebooks for school, my wallet, a book of poems I'd been reading, my journal.

My journal must have been what they were looking for, because Meredith's British girlfriends testified after my arrest that I'd been writing in it in the waiting room at the *questura*. I had done so to calm myself, but soon the contents were leaked to the press. In it, they found, among other things, my comments about wanting to compose a song in tribute to Meredith. (Ironically, I would later get a bill for the translation of the journal into Italian.)

The police officer who retrieved my things that day was the same one who usually came to the prison when the prosecution wanted to confiscate my belongings or have me sign a document about forensic analysis—an unshaven, overweight man with a crew cut. He was the cop who, during the interrogation, thought I'd told him, "Fuck you!" and who yelled it back at me.

He asked if I'd seen the news about the murder weapon.

I glared at him. "It's a mistake," I said. "I wasn't anywhere near Meredith when she was murdered and neither was Raffaele's knife."

The officer shook his head and laughed derisively. "Another story? Another lie?" he scoffed. He looked at me as if I were the most vile, worthless thing he'd ever laid eyes on. No one had ever stared at me with so much hatred. To him, I was a lying, remorseless murderer. I heaved back great waves of anger but waited to get back to my cell before I broke down at the ugliness of it all—my friend being dead, my being in prison, the police following a cold and irrational trail because they had nothing better.

Chapter 18
———————

November 2007

D uring my first few days in prison, I would have loved
any distraction, but the TV was covered up and
declared off-limits. When I moved in with Gufa, I went
from no news to a barrage. The TV blared practically 24/7.

Lupa, the *agente* who helped me so much in my early time at
Capanne, had cautioned me: "The media are saying horrible
things about you. Don't pay attention to it," she advised. "It will
just upset you."

She was right.

My Italian was still elementary enough that if I wasn't paying
close attention, I couldn't grasp much of what was being said. I
embraced my new routine—do as many sit-ups as I could man-
age, write, read, repeat—as if ignoring the reports would make
me immune to them, that they couldn't hurt me. I convinced
myself that whatever awful things the media were saying about
me were irrelevant to the case. *It doesn't matter*, I told myself. But
in my heart I knew it did.

Mentally tuning out the TV helped, but it was impossible to
drown out all the coverage—there was a television set bolted to

the wall in every cell, and the set and I were locked in the same room twenty-three hours a day. The screen was flooded with my image. I felt as though I were looking at someone else. A picture of me talking to detectives outside the villa was a news channel staple. They replayed the footage again and again, often in slow motion, of Raffaele and me kissing in the villa's front yard after Meredith's body was found. *They're making it into something it wasn't. The way they're manipulating this, people will think Amanda just couldn't keep her lips off Raffaele.* They acted as though our affection showed such a flagrant disregard for Meredith that it was obvious Raffaele and I were hiding the truth. The commentators pointed to our consoling kisses as proof that we were capable of murder. Their remarks were so unfair, their expressions so smug. I wanted to scream, "Look into our faces! Do we really look ready to jump on each other and have sex in the driveway?" What I saw then—and see now—is a young girl and guy in shock.

I felt violated, indignant that journalists could say or imply anything they wanted, that they could use my photo as a symbol of evil. I now understood the belief in some tribal cultures that having your picture taken robs you of your soul.

Reporters also plundered my Myspace page, and this felt just as intrusive.

When I created my social networking profile in high school, borrowing the soccer moniker my teammates had given me when I was thirteen seemed safer than using my real name. Sure, I knew *foxy* meant "sexy" or "sassy," but that was the irony of it—and the fun. My soccer girlfriends had ironic and sassy nicknames, too. Martinez was Martini; Miller was Miller Light; Trisha was Trash. By college, when I graduated to Facebook, I seldom looked at Myspace. I could never have dreamed that something so harmless could later have such damning results, that the prosecution would focus on my nickname's other meanings—"wily" or "tricky."

Overnight my old nickname became my new persona. I was now known to the world as Foxy Knoxy or, in Italian, *Volpe Cattiva*—literally, "Wicked Fox."

"Foxy Knoxy" was necessary to the prosecution's case. A regular, friendly, quirky schoolgirl couldn't have committed these crimes. A wicked fox would be easier to convict.

They were convinced that Meredith had been raped—they'd found her lying on the floor half undressed, a pillow beneath her hips—and that the sexual violence had escalated to homicidal violence.

They theorized that the break-in was faked.

To make me someone whom a jury would see as capable of orchestrating the rape and murder of my friend, they had to portray me as a sexually deviant, volatile, hate-filled, amoral, psychopathic killer. So they called me Foxy Knoxy. That innocent nickname summed up all their ideas about me.

"Foxy Knoxy" also helped sell newspapers. The tabloids mined my Myspace profile and drew the most salacious conclusions. I resented that they took my posts and pictures out of context, emphasizing only the negative. A photo of me dressed in black and reclining provocatively on a piano bench, a shot my sister Deanna had taken for a high school photography class, circulated. They published parts of a short story I'd written for a UW creative writing class, about an older brother angrily confronting his younger brother for raping a woman. The media read a lot into that. There were pictures of me at parties and in the company of male friends, and a video showing me drunk.

These were snippets of my teenage and college years. Not shown were the pictures of me riding my bike, opening Christmas presents, playing soccer, performing onstage in my high school's production of *The Sound of Music*. Looked at together, these latter images would have portrayed a typical American girl, not as tame as some, not as experimental as many, but typical among my age group—a group that had the bad judgment to put our lives online. Now, at twenty, all I could think was, *Who's writing these articles? Is no one being fair?*

"Foxy Knoxy, the Girl Who Had to Compete with Her Own Mother for Men" ran in England's *Daily Mail*. The writer speculated that my mom's marriage to Chris, a man they described as "young enough to be [my] own brother," intensified my feelings of rejection that "culminated" in Meredith's death. They conveniently overlooked the part of my Myspace page that read, "Foxy Knoxy's heroes." My answer was: "My mom."

My supposedly obsessive promiscuity generated countless articles in three countries, much of it based on information the police fed to the press. It seemed that the prosecutor's office released whatever they could to bolster their theory of a sex game gone wrong. They provided descriptions of Raffaele's and my public displays of affection at the *questura* and witness statements that portrayed me as a girl who brought home strange men. Whatever the sources, the details made for a juicy story: attractive college students, sex, violence, mystery.

I became the embodiment of everyone's worst fears of, or fantasies about, a sexually aggressive woman. I could't deny that I'd hooked up with a couple of guys in Perugia whom I hadn't known well. But I hadn't sought out men because I was obsessed with sex. I was experimenting with my sexuality. My reaction to being characterized as a femme fatale was *Me? Really? Of all people!*

Along with my exaggerated sexual history, people found it tantalizing that I didn't look like a depraved murderer. The press said that I had "the face of an angel but the eyes of a killer" or "an angel's face and a demon's soul." Suddenly I had a "secret side."

Soon after I got to Capanne, I started getting fan mail—some from people who thought I was innocent, and some from strangers who said they were in love with me. I appreciated the encouraging letters and was shocked, and baffled, by the others. It seemed to me that these men—often prisoners themselves—had written me by mistake. Their passionate, sometimes pornographic scribbling had nothing to do with me and everything to do with the media's creepy, hypersexual creation. I'd never imagined that I would be bombarded with such perverted attention. And if I was drop-dead sexy, it was news to me.

Vice-Comandante Argirò always made a production out of opening my mail, winking and chattering about how many admirers I had. But I got at least as much hate mail as I did supportive and provocative letters combined. Some of it terrified me, especially the chicken-scratch notes with no return address that said they knew where my parents were staying and planned to cut up their faces. What if they followed through? I warned Mom to be careful, to close her windows at night. And after one particularly threatening letter, I told Argirò, hoping he'd alert my mom right away. "Forget about it," he said, dismissively. "They're just words."

I felt terrible that my mom and dad had abandoned their regular lives to come to Italy, and that their spouses back home were being hounded by journalists and paparazzi, who staked out their houses, waiting for them to come or go, knocking on the door and phoning them incessantly. The people who hated me couldn't get to me in prison, but there was nothing I could do to protect my family.

Nor could I deflect the overpowering attention being given to "the Wicked Fox." The real me had been lost. It seemed as though people were putting me in a costume that trapped me even more than the iron bars I lived behind.

The portrait the prosecution and media created of me as sex-charged led some people to make sure that everything Raffaele and I had done in the days following the murder, every errand we ran, appeared sexual. Soon after the first few Foxy Knoxy stories showed up, the owner of Bubble, the cheap teen shop where Raffaele and I had stopped the night after Meredith's body was found, told journalists that I'd bought a red G-string. Stories ran under headlines such as "Pictures of the Moment Foxy Knoxy Went Shopping for Sexy Lingerie the Day After Meredith's Murder," quoting Raffaele supposedly saying, "I'm going to take you home so we can have wild sex together."

As usual Luciano and Carlo filled me in on this story. "But I didn't buy sexy underwear!" I protested. "And Raffaele didn't say that. It was red, but it's a pair of bikini briefs with a cartoon cow on it. I was locked out of my house and had only the clothes I was wearing."

"I'm sorry to even ask you about it, Amanda," Carlo said gently. "We just needed to know what to make of the claim."

Carlo and Luciano urged my family and me to ignore the media. They organized a short press conference at which my parents read a statement saying that I was innocent. After that, the lawyers refused to allow my family to answer any questions from journalists. We'd learned that anything could be turned around and used against us.

"The media are about as evil as you can get," Carlo would say. "They're going to do whatever makes money. Anyone who meets you will see you're not the girl the prosecution and press are portraying, but journalists aren't interested in hearing that you're a good girl. We have to do that in the courtroom. Don't worry. We'll have our chance."

I didn't know that my parents were debating this approach with the lawyers. Mom and Dad wanted to stand up to the media. They understood that once damaging words are unleashed, they stick. To protect one another from added pain, there was a lot Mom, Dad, and I didn't say during our visits.

Besides all the lies about my out-of-control sex life, people started pitting Meredith and me against each other. It had never been that way when she was alive. Meredith's British girlfriend Robyn Butterworth gave a witness statement after my arrest claiming that Meredith had complained about my loud singing and poor toilet hygiene. Police leaked parts of the testimony to the press, and like so many other things, normal moments in the lives of housemates were refashioned into a motive for murder. I do sing loudly and often. And I knew Meredith had been embarrassed to tell me that the toilet needed to be brushed after each use. I'm embarrassed to think that she may have put off bringing it up with me until it happened a few times. Still, I thought she probably hadn't complained as much as mentioned it to friends or family to ask how to handle it so she wouldn't hurt my feelings.

The idea that Meredith and I had been at odds ramped up quickly in the press. A couple of weeks after Robyn's statement came out, investigators announced they'd found my blood on the faucet in the bathroom that Meredith and I had shared. Prosecutor Mignini hypothesized that the two of us had gotten into a fistfight and I'd wound up with a bloody nose. The truth was far less dramatic—and less interesting. I'd just gotten multiple piercings in both ears, and I took out all eleven earrings so that I could wipe my ears each morning while the shower water heated up. When I noticed the tiny droplets of blood in the sink the day Meredith's body was discovered, I thought the blood had come from my ears, as it had on another day, until I scratched the porcelain and realized the blood was dry. That must have been what was on the faucet.

Meredith had been dead for just three weeks. I still could barely process the loss of my friend. It infuriated me that the media were rewriting our relationship to fit their storyline. I was a monster. Meredith was a saint. The truth was that we were very much alike. She was more contained than I was, but we were both young girls who studied seriously and wanted to do well, who wanted to make friends, and who'd had a few casual sexual relationships.

Raffaele didn't demonize me, but he did publicly renounce me. Answering questions through his lawyer, he told one journalist, "If I am here it's her fault above all. I am conscious that contrary to what I thought, our paths have diverged profoundly."

When asked what he'd like to say to me, his answer was "Nothing. I have absolutely nothing to say to her."

I didn't know what to think about Raffaele. Hearing that he'd destroyed my alibi was as baffling as it was incensing. Saying I'd put him up to lying was inexcusable and painful. *And now this*, I thought. *Did I misjudge him?* I didn't think so, but I wasn't at all sure what to make of him. One day we were really close, and the next he announced that he'd dropped me. Had this come from him? His lawyers? Journalists? I rationalized that I wasn't the Italian girl he needed. I tried to be forgiving. *If Raffaele doesn't want to talk to me again, I'll understand. This has been traumatic for everyone.* But sometimes I was just angry.

I was nursing these hurts when I got news so shattering that it blotted out almost everything else. I was at my nightly infirmary appointment, where I was meeting with a doctor I'd never seen before. Dressed in a white lab coat, my medical file in hand, he said, "We got the results of your blood test." His bedside manner was as warm as gelato. "You tested positive for HIV."

I was so shocked I couldn't think. I couldn't make sense of what he was saying.

The doctor saw my panic. "Don't worry," he said, offering me a spoonful of compassion. "It could be a mistake. We'll need to do more tests."

His reassurance struck me as hollow, as if he were just trying to postpone my inevitable anguish. I thought my head would explode from anxiety. I was in prison for a crime I hadn't committed, and now I might be infected with HIV?

Argirò was standing a foot behind me when I got the news. "Maybe you should have thought about that before you slept with lots of people," he chided.

I spun around. "I didn't have sex with anyone who had AIDS," I snapped, though it was possible that one of the men I'd hooked up with, or even Raffaele, was HIV-positive.

"You should think about who you slept with and who you got it from."

Maybe he was trying to comfort me or to make a joke, or maybe he saw an opening he thought he could use to his advantage. Whatever the reason, as we were walking back upstairs to my cell, Argirò said, "Don't worry. I'd still have sex with you right now. Promise me you'll have sex with me."

I was too undone to react.

Sitting on my bed, I wondered if I would die in prison. I didn't know then that people live with HIV for a long time due to improved meds. *Please, please, let it be a mistake. Please let it be wrong. I don't want to die. I want to get married and have children. I want to be able to grow old. I want my time. I want my life.*

I didn't know how to tell Mom or Dad. I desperately wanted to talk to them, but their next visit wasn't for two more days. I miserably reasoned that I'd had such a fortunate life that all my bad luck was catching up to me now.

I was aware that there were consequences to being careless about sex. I thought I'd been careful enough. But what had I really known about my sexual partners? Why hadn't I seriously considered the risk? I'd been trapped by prison; now I felt trapped by my own body, trapped by my stupidity, trapped because bad things happened to people for no reason, with no way of anticipating them. Thinking about the life I might have had instead of the one I was living made me understand for the first time how people in mourning tear their clothes or rip out handfuls of hair. I wanted to undo everything—to be out of my body, out of this prison, out of this life that had caved in on me. I buried my face in my pillow so no one could hear me and wailed.

So much had happened that I didn't know how to handle emotionally or practically: Meredith's death; my interrogation, arrest, and imprisonment; HIV. Any one of them would have been a hard burden for a twenty-year-old. To have them all at once was devastating. Every problem put before me was foreign, and the tools I had—stubbornness, optimism, the support of my family, and the certainty of my innocence—weren't nearly enough for the situation.

Part of me couldn't believe I really had HIV. Even though the media were portraying me as a whore, I knew I wasn't one. It seemed too ironic, too overwhelming that all this was happening at once. *Just breathe. Write down that you're freaking out and then stop. You're not going to make anything better by going crazy over it. Relax. The doctor said they don't know that you have it for sure.*

I got out my diary to think this over rationally, imagining who could have infected me, replaying my sexual experiences in my mind to see where I could have slipped up. I wondered if a condom had broken, and if so, whose. If it had, did he know?

I'd had sex with seven guys—four in Seattle and three in Italy. I tried to be logical, writing down the name of each person I'd slept with and the protection we'd used.

Writing made me feel a little better. I knew I needed to get out of prison and get checked by someone I trusted before I started thinking and acting as if my life were over. I forced myself not to anticipate the worst.

That Saturday, I told my parents what the doctor had said. My mom started crying immediately. "But I haven't had unprotected sex," I said, trying to reassure her. "I'm sure it's going to be fine."

My dad was skeptical. He asked, "Do you even think they're telling you the truth?"

That possibility hadn't occurred to me. But when I told them, Luciano and Carlo seconded that idea. "It could be a ploy by the prosecution to scare you into an even more vulnerable emotional state so they can take advantage of you," Carlo said. "You need to stay alert, Amanda, and don't let anyone bully you."

In the end, I don't know if they made up the HIV diagnosis. It wasn't the doctor who said I should think about whom I'd had sex with, but Argirò. It might have been that the test was faulty, or Argirò could have put the medical staff up to it so he could ask me questions and pass the answers along to the police.

It was nearly two months before the doctors let me know that the HIV test had come out negative. When they did, I thought, *Oh, thank God!* But I was still seeing the doctors twice a day, and it had been a long time since anyone had even brought it up. The possibility no longer scared me as much, and I'd begun to assume everything was okay.

A week after I got the original HIV news, a guard took me down to the offices on the main floor, where three police officers were waiting for me. "We have a warrant to search your cell," they said. "We'll give you a five-minute head start to destroy whatever you'd like, or you can let us go up immediately."

"You can come now and look through whatever you want," I said.

I wondered what they were hoping to find. Did they want to search my clothing for traces of Meredith's blood? I felt almost smug, because I knew they wouldn't find anything incriminating, and I hoped it might convince them that I truly had nothing to hide.

The cops spread out all my papers and documents on my bed. They confiscated anything with my handwriting on it—my grammar exercises, unfinished letters, notes, my prison diary—and left everything else. That's when I understood. They wanted to see what I was thinking.

The physical chaos they left behind was nothing compared to

the chaos in my head. They'd penetrated my innermost space, demonstrating to me that nothing was safe from them.

A few months after that, they released my prison journal to the media, where instead of reporting that I'd had seven lovers altogether, some newspapers wrote that Foxy Knoxy had slept with seven men in her six weeks in Perugia.

Chapter 19

November 18–29, 2007

I was stunned one morning when I looked up at the TV and noticed a breaking news report. There was now a fourth suspect, and an international manhunt for him had been launched.

The police didn't say who the suspect was or how this person fit into the murder scenario they'd imagined, only that they'd found a bloody handprint on Meredith's pillowcase that wasn't mine, Patrick's, or Raffaele's.

The news rattled me, but it also gave me hope. Maybe this meant the police hadn't completely given up trying to find the truth. For the next twenty-four hours I was consumed by the question *Who is this unnamed person?*

I, and everyone else watching TV, found out the next day. His name was Rudy Guede. The police had his fingerprints on file because he was an immigrant with a green card.

The name didn't click until I saw his mug shot.

Oh my God, it's him.

I thought back to November 5, when I was sitting in the hall at the *questura*, assuming I was just waiting for Raffaele, and talking to the silver-haired cop. As I'd been doing for days, I was trying to recall all the men who had ever visited our villa, when I suddenly remembered one of Giacomo and Marco's friends. It had annoyed me that I couldn't remember his name. "I think he's South African," I told the detective. "All I know is that he played basketball with the guys downstairs. They introduced him to Meredith and me in Piazza IV Novembre in mid-October. We all walked to the villa together, and then Meredith and I went to their apartment for a few minutes."

I'd seen Guede just one time after that. He'd shown up at Le Chic, and I had taken his drink order. Those few words were the only ones we ever exchanged.

I was still living in semi-isolation, meaning that I wasn't allowed to participate in group activities or speak to other prisoners. But when I'd asked to be moved from Gufa's cell, I'd really hoped that meant they would put me by myself again. Instead I'd been moved into a cell with three older women. And just as with Gufa, the TV in Cell No. 10 was on all the time.

The only difference was that with the announcement about Guede, now I couldn't watch the news enough.

I learned that Guede was twenty and originally from Ivory Coast. He'd been abandoned by his parents and taken in by a rich Perugian family who treated him like a son. He was a talented basketball player who'd made a lot of friends on the court. But over time, he'd been more inclined to loaf than to work, and his surrogate family disowned him. He'd lost his job in the fall of 2007, before Meredith and I met him. Guede had been caught breaking into offices and homes and stealing electronics and cash.

Another report said that in mid-October he'd thrown a large rock through a window at a Perugian law office to get inside. *A broken window and a rock on the floor? Exactly what we'd found in Filomena's room.* He'd stolen a laptop and a cell phone from the firm.

I couldn't believe that none of us had picked up on how shady Guede was. Just a few days before Meredith was killed, the director of a Milan kindergarten arrived in the morning and caught Guede coming out of her office. When the police got there, they found one of the kindergarten's kitchen knives in his backpack, along with the laptop from the law office, a set of keys, a woman's gold watch, and a small hammer he'd used to break glass. The police were on the verge of arresting him for that crime but released him without giving a reason. I couldn't understand how they'd let Guede slip through their fingers. All I could think was that if he'd been put behind bars then, Meredith would still be alive.

It didn't make sense to me that they had let him go but had leapt to arrest me.

I'd met but didn't know Rudy Guede. I didn't know if he was capable of murder. I couldn't imagine why he might do something so brutal. But I believed that he was guilty, that the evidence could only be interpreted one way. Finally the police could stop using me as the scapegoat for some phantom killer whom no one could name—a phantom whose place I'd been filling.

For nearly three weeks I'd been unable to think of anyone, however distant, who could have stabbed Meredith to death. Now there was a face and a name. It was awful, but it was a relief.

Still, I was surprised it was Guede who had been named,

because the two times I'd met him were under such ordinary circumstances. There was nothing distinguishable about him. He'd seemed interchangeable with almost every guy I'd met in Perugia—confident, bordering on arrogant. Not threatening. Not like a down-and-out thief. Not even odd.

The next day the same police officer who'd mocked my reaction to the DNA evidence the prosecution claimed was found on the knife brought documents to Capanne for me to sign. This happened regularly during the investigation phase—they had to notify me whenever they confiscated anything from the villa, analyzed forensic evidence that pertained to me, or, unbelievably, were billing me for investigative expenses. I became used to the bureaucracy. But I was never prepared for the cop's cruelty. He was talking so fast that I caught only one word: "Rudy."

"Rudy?" I asked, repeating his name to make sure I'd heard correctly. "You mean the guy who police are calling 'the fourth person'?"

"Yes, Rudy. You know him?"

"Vaguely," I answered, shrugging.

"Vaguely, huh? We'll see what he says about that," the cop said.

I didn't respond but tried to act confident so he wouldn't think he was getting to me. I was thinking, *Guede won't have anything to say about me. He doesn't know me.*

On November 20, German police found Guede in Germany, where he'd fled on November 3, the day after Meredith's body was discovered. He was riding a train without a ticket when he was picked up and taken into custody as a murder suspect.

Within hours, I learned that, before his arrest, he told a friend over Skype, as Perugian detectives listened in, that he'd been at the villa the night of the murder. "I was in the bathroom when it happened," he said. "I tried to intervene, but I wasn't able. Amanda has nothing to do with this . . . I fought with a male, and she wasn't there." Neither was Patrick, he said. "The guy was Italian, because we insulted each other and he didn't have a foreign accent."

When his friend asked if it was Raffaele, "the one from TV," Guede said, "I think so, but I'm not sure."

After his arrest, Guede told German police that Meredith had invited him to meet her at the villa, and that they'd been fooling around when he felt sick from a kabob he'd eaten earlier. He said he was in the bathroom listening to his iPod when he heard Meredith scream. A brown-haired Italian man he couldn't identify committed the murder. Guede had tried to help Meredith as she was dying, staunching the blood with towels, but fled when he realized there was nothing he could do. He said he was afraid that because he was black, he'd be condemned for a crime he hadn't committed.

Guede apparently tried to establish an alibi by changing clothes and heading to a downtown dance club hours after the murder. His lawyers later said he'd been so frightened by the murder that he'd gone there to calm himself down. He went to Domus again the next night—attracting attention when he continued dancing during a moment of silence for Meredith. He left town the following day. Carlo and Luciano told me he probably got spooked by the media's attention to the case and decided it was best to leave and take his bloody clothes and shoes with him. They guessed that Guede had probably been in the middle of robbing the villa when Meredith came home, and he had attacked her. As soon as they suggested this scenario, it made perfect sense to me.

I hadn't been able to put all those pieces together before. Meredith's murder had been so horrific, and my arrest too absurd, it had been impossible for me to think logically about it.

I saw it as a momentary problem that Guede was fingering Raffaele, but this was huge! Guede had backed up my alibi: I hadn't been at the villa. And since I hadn't been there, since I'd been at Raffaele's apartment, Raffaele would be cleared, too. We would both be freed.

Seeing how the prosecution treated Patrick in the two weeks since his arrest should have given me insight into how they worked. My lawyers told me it had been widely reported the week before that Patrick had cash register receipts and multiple witnesses vouching for his whereabouts on the night of November 1. A Swiss professor had testified that he'd been at Le Chic with Patrick that night from 8 P.M. to 10 P.M. But even though Patrick had an ironclad alibi and there was no evidence to prove that he'd been at the villa, much less in Meredith's bedroom at the time of the murder, the police couldn't bear to admit they were wrong.

Patrick went free the day Guede was arrested. Timing his release to coincide with Guede's arrest, the prosecution diverted attention from their mistake. They let him go only when they had Guede to take his place.

Watching the footage of Patrick walking out of prison and standing with his wife and baby, I flashed back to the awful hours in the middle of the night on November 6 when I was being interrogated: I was weak and terrified that the police would carry out their threats to put me in prison for thirty years, so I broke down and spoke the words they convinced me to say. I said, "Patrick—it was Patrick."

I dreamed about the interrogation almost every night during these early days in prison. I would be back in the crowded, close interrogation room, feeling the tension, hearing the officers yelling, reliving the primal panic. I'd wake up sweating, my heart banging. Nothing in my life up to then had compared to that experience. What had happened to me that night? How I could I ever have named Patrick?

As I watched his release, I felt an enormous emotional burden lift from my heart. Justice had been done. The police had cleared him of any wrongdoing. He would no longer have to suffer from my irrational mistake. I clapped my hands together almost gleefully.

Then I immediately felt embarrassed, self-conscious that, in one way or another, the few prisoners and guards who happened to see this would misread my actions as selfish. I didn't know whether the guards were reporting directly to the prosecution, but I knew that everyone thought I was a liar and that anything I said and did would be viewed from that angle—that I was trying to make people think I was innocent by acting happy for Patrick. The police would almost certainly think this was one more instance of Amanda Knox behaving inappropriately—one more example of me as a manipulative, depraved person.

Even if my cellmates didn't see my reaction as putting on an act, I didn't want anyone to know what I was actually thinking and feeling. I was protective of myself in that environment. I felt vulnerable and scared, and I didn't want anyone to see that, even if that's how I really felt.

In truth, I did see Patrick's release as my vindication. By writing my two postinterrogation statements—my *memoriali*—I had tried to convince the police that Patrick was not Meredith's murderer. And now the prosecution knew that when I retracted my declarations from that night, I was telling the truth: Patrick *was* innocent. Raffaele and I *had* been together at his apartment the whole time.

Obviously Mom, Dad, and I wouldn't make it to Seattle for Thanksgiving—which was just two days away—but I did think I would get out of prison and be allowed to stay with my parents while the investigation continued. Christmas in Seattle seemed likely.

The prosecution would understand how, under pressure during my interrogation, I had pictured a scene that wasn't true. I had faith that my lawyers could prove the knife with Meredith's and my DNA was a mistake. My confidence was bolstered by Guede's arrest. I didn't know him. If he was Meredith's murderer, I was sure people would see that Raffaele and I had had nothing to do with it.

Soon I'd be cleared as a suspect.

When Carlo and Luciano came for their next scheduled meeting I was happier than I'd been since before Meredith's murder. We sat in the office where we met each week. Luciano held my hand while Carlo said, "The prosecution has no intention of releasing you, Amanda. They're just subbing Guede in for Patrick."

The prosecution could have redeemed themselves. Instead, they held on to Raffaele and me as their trophies.

I learned that when he signed the warrant for Patrick's release, Giuliano Mignini said that I'd named Patrick to cover up for Guede. It was his way of saying that the police had been justified in their arrest of three people and that any confusion over *which* three people was my fault. I was made out to be a psychotic killer capable of manipulating the police until my lies, and the law, had caught up with me.

Patrick gave only one interview condemning the police for his unfounded arrest before his lawyer, Carlo Pacelli, advised him to side with the prosecution, who had taken him away in handcuffs, humiliating him in front of his family, in the intimate hours of the morning. After that, he announced that he would never forgive me for what I had done, that I'd ruined him financially and emotionally. He talked about my behavior in his bar, saying that he'd fired me for flirting with his customers. He called me "a lion," "a liar," and "a racist."

The truth is that he had hired me not just to serve cocktails but to bring in customers. He had cut back on my days because I was a mediocre waitress and not enough of a flirt to add to his bottom line. Then, after Meredith's murder, I quit because I was afraid to be out alone at night.

I absolutely understood why he was angry with me. I'd put his reputation, his livelihood, and possibly even his life at risk. I felt sick with guilt. I thought he deserved an explanation and an apology from me. When I asked my lawyers if it would be okay for me to write him, they shook their heads no. "I'm afraid it's not as simple as that anymore," Carlo explained. "Patrick's lawyer will hand over anything you send Patrick to the press."

Any communication with Patrick would be publicized and scrutinized and played to my disadvantage, especially if I explained why I'd said his name during my interrogation. I'd have to go into how the police had pressured me, which would only complicate my already poor standing with the prosecution. If I said I'd imagined things during the interrogation, I'd be called crazy. If I said I'd been abused, it would be seen as further proof that I was a liar.

I know my lawyers' interest was to protect me from the prosecution and the media, but I wished then and I wish today that I'd taken the risk of writing to Patrick anyway. I owed him that.

———————

Chapter 20

December 2007

When I first told Carlo and Luciano I wanted to talk to Prosecutor Mignini, I didn't think of it as a rematch between opposing sides. I saw it as a chance to set the record straight. Finally.

"I'm sure if I talk to him in person, I can show him I'm sincere," I told my lawyers. "I can convince him he's been wrong about me. It bothers me that everyone—the prosecutor, the police, the press, the public—thinks I'm a murderer. If I just had the chance to present my real self to Mignini I'm sure I could change that perception. People could no longer say I'm a killer."

Carlo and Luciano looked at me doubtfully. "I'm not sure it's the best idea," Carlo said. "Mignini is cagey. He'll do everything he can to trick you."

"I feel like it's my only hope," I said. "My *memoriali* didn't change anyone's mind—they just made the prosecution and the media portray me as a liar. I didn't get to tell the judge what happened before she confirmed my arrest. I think I have to explain face-to-face why I named Patrick. I've got to make Mignini understand why I said I'd met Patrick at the basketball court, why I said I'd heard Meredith scream."

"He can be intimidating," Carlo said.

"The thought of meeting with Mignini makes me incredibly anxious," I said. "I know what it's like to be bullied by him. But I have to try."

My thought was that I had misled the police. I needed to take responsibility for my mistake. It seemed like the right, and adult, thing to do.

"Nothing good is going to come of this," Luciano grumbled.

But when my lawyers came to Capanne the following week, they told me that they'd decided, reluctantly, to arrange a second interrogation.

"It's risky," Carlo said. "Mignini will try to pin things on you."

"He already has," I told them.

The first time I met Mignini at the *questura*, I hadn't understood who he was, what was going on, what was wrong, why people were yelling at me, why I couldn't remember anything. I thought he was someone who could help me (the mayor), not the person who would sign my arrest warrant and put me behind bars.

This time I was ready. This time my lawyers would be there. I'd be rested. My mind was clear. I was going in knowing what I was getting into. I'd take my time and answer all his questions in English. I didn't think I'd be released immediately, but I hoped that giving the prosecutor a clear understanding of what had happened would help me. Then, as new evidence came forward proving my innocence, Mignini would have to let me go.

I now had a standing Wednesday morning appointment with my lawyers. Each week, as I walked into the prison office that doubled as our meeting room, both men would stand up and say, "*Ciao*, Amanda." Then Luciano would tilt his head back, look up at the ceiling, and say, "*Ciao, polizia*," before he'd look back at me: "*Teniamoci conto degli altri ospiti alla festa*"—"Let's keep in mind the other guests at the party."

Luciano's jaunty greeting was not my only clue that the room was almost certainly bugged. My lawyers had been clear: "Never repeat anything about your case to anyone," Carlo said. "I'm sure you're being watched and listened to. I understand your need to talk freely with your parents, but the police will take advantage of anything they can to build their case against you. Please be careful."

"Okay," I said.

But I wasn't good at censoring myself. I had only two hours a week with my mom and dad, and they were the only people I could open up to. It made me feel better to vent, and my parents needed to know what I was thinking. I couldn't see the danger in discussing with them my day-to-day prison life, my interactions with my cellmates and guards, or my case. Since I hadn't been involved in the murder, I figured that anything I said would only help prove my innocence.

I hadn't considered that the prosecution would twist my words. I didn't think they would be capable of taking anything I said and turning it into something incriminating, because everything I said was about my innocence and how I wanted to go home. I was saying the same thing again and again.

On their first visit after the knife story came out, Dad and Mom were telling me my lawyers' theory—that the police could be using the knife as a scare tactic to get me to incriminate myself. "The police have nothing at all on you," Mom said. "So they are trying . . . to see if you['ll] say something more."

"It's stupid," I said. "I can't say anything but the truth, because I know I was there. I mean, I can't lie about this, there is no reason to do it."

What I meant by "I was there" was that I was at Raffaele's apartment the night of Meredith's murder, that I couldn't possibly implicate myself. I hadn't been at the villa. I wasn't going to slip up, because I wasn't hiding anything.

Sitting next to me at the table in the room where we'd been reunited a few weeks earlier—the same room where I met with my lawyers—Mom held my hands in hers, nodding in agreement at me. Then we moved on to other topics, such as how we each were getting through this and what friends and family at home were doing to try to help.

The police did not move on. They seized on my comment, which they had on tape. A couple of weeks later, in early December, a convoluted version of what I'd said made international headlines, including the London *Telegraph*'s "Tape 'Puts Knox at Meredith's Murder' Scene."

The article began, "Dramatic new evidence has emerged that may help prove that Amanda Knox, the American girl accused of murdering Meredith Kercher, was present when the British student died."

The police had leaked the false but enticing tale to the press.

Luciano and Carlo understood what I hadn't yet grasped: that the prosecution was so fixated on proving my guilt, they saw only what they wanted to see, heard only what they wanted to hear, found only what they wanted to find. Facts be damned.

I was indignant. "How can they do that?" I asked. "It's straight-up false!"

"Don't worry," Carlo said. "We'll be able to prove it's wrong once the prosecution gives us the transcripts. But please use this as a lesson, Amanda. The prosecution will pounce on anything they believe will serve their purposes. Please remember the room where you and your parents visit is bugged."

Being more careful in the future wouldn't immediately resolve this serious misunderstanding. A few days later the judge considered those words when deciding if I could be moved to house arrest. In another crushing blow that characterized my early months in prison, my request was denied. I was stuck alone behind bars.

Calling the intercepted conversation a "clue," the judge wrote, "it can certainly be read as a confirmation of the girl's presence in her home at the moment of the crime."

He went on to describe me as "crafty and cunning," saying that I was "a multifaced personality, unattached to reality with an elevated . . . fatal, capacity to kill again."

It wasn't until my pretrial, the following September, that a different judge agreed with my defense that it was obvious I was talking about Raffaele's apartment, not the villa, and removed this "evidence" from the record.

Just as Carlo had told me not to discuss my case, he'd also warned me to write down as little as possible, caution that I thought was borderline paranoid. I'd started keeping a journal as soon as I learned to write complete sentences, and I didn't see why I should stop now, when I needed that outlet the most. Even after my prison diary was confiscated, I didn't worry about anything I'd written. I wasn't guilty. I didn't think about what could happen once my words were out of my hands.

Not even my lawyers understood my journal musings on Raffaele and the knife that made their way into the newspapers. I'd written a hyperbolic explanation about him taking the knife from his apartment behind my back. I had to explain to Carlo and Luciano that I'd concocted it because the possibility of a knife with Meredith's DNA coming out of Raffaele's apartment had struck me as so preposterous:

> *Unless Raffaele decided to get up after I fell asleep, grabbed said knife, went over to my house, used it to kill Meredith, came home, cleaned it off, rubbed my fingerprints all over it, put it away, then tucked himself back into bed, and then pretended really well the next couple of days, well, I just highly doubt all of that.*

But I didn't have the luxury of explaining what I'd written to everyone who read it. After my passage was translated into Italian and then retranslated back into English, it bore little resemblance to the original—and a great resemblance to the prosecution's theories about what had happened the night of November 1:

That night I smoked a lot of marijuana and I fell asleep at my boyfriend's house. I don't remember anything. But I think it's possible that Raffaele went to Meredith's house, raped her and then killed her. And then when he got home, while I was sleeping, he put my fingerprints on the knife. But I don't understand why Raffaele would do that.

Once I had my meeting with the prosecutor I'd correct all the confusion about me. I thought my upcoming interrogation would tie up all these loose strands.

Carlo and Luciano warned me once again that it might not be so simple. "Mignini will ask pointed questions to snare you," Carlo said, his face serious. "He will try to paint you as a liar. He wants to show that you have a connection to Rudy Guede. He'll try to prove that you lied about Patrick on purpose. Are you prepared for that, Amanda?"

"I know," I said. "I'm ready."

But I didn't—and I wasn't.

As the date for the interrogation approached, Luciano and Carlo offered me a few pointers. "Don't let him get to you. Don't say anything if you don't remember it perfectly. It's okay to say, 'I don't remember.' You don't have to be God and know everything. It's better to say, 'I don't know,' and move on."

I was a jumble of emotions—eager to set the prosecutor and the public straight on who I really was and nervous about putting myself out there. But the night before Interrogation Day, my nerves overtook my excitement. I couldn't eat much of the pizza my roommates and I made for dinner on our camp stove. I turned and tossed most of the night, thinking about what I wanted to tell the prosecutor. As I was being escorted to the prison compound's center building at 10 A.M. the next day, I was humming my prison anthem, "Let It Be," trying to calm some of my jitters.

The meeting took place in the same makeshift courtroom as my hearing to confirm my arrest five weeks earlier. The setting wasn't that much more pleasant than the *questura* office where Mignini had interrogated me the first time. Separate tables for the defense and the prosecution faced each other from opposite sides of the small, dim, bare room, with two barred windows set close to the ceiling so no one could see in or out.

The tension was instantly obvious. Mignini was sitting at his table with two police officers. Like Carlo and Luciano, he was wearing a black robe. The three men had come ready for a fight. I felt awkward and out of place, as though I'd stepped into the middle of a feud that had nothing to do with me.

But I was the reason for the feud—and the only person who could set things right.

I stood near Carlo and Luciano with an interpreter, waiting for Mignini to give me permission to speak. That never came. Instead of asking what I had to say, he started firing questions at me immediately.

What has stuck with me the most is that he never looked me in the eye. He stared down at the paper in his hand, on which his questions were written out. It's as if I didn't merit the effort it would have taken to look up.

"Do you have any Spanish friends?" he asked—Rudy Guede said he hung out with Spanish friends on Halloween.

I was calm and assertive. "No," I answered.

"What's the meaning behind your name Foxy Knoxy?"

"It's just a nickname," I said.

"But what is the meaning behind it?"

"There is no meaning behind it. It's a play on my last name, Knox. My soccer teammates started calling me that as a teenager."

"Why do you use it to identify yourself?"

"I don't. I don't introduce myself as 'Foxy Knoxy.'"

"Did you have problems with Meredith?"

"No. We didn't know each other long, but we were friends."

"Do you know Rudy Guede?"

"I met him," I said, "but I didn't remember his name until he was arrested."

Mignini grilled me about my drug use, the people I knew in Perugia, the friends I'd invited over to the villa. He asked me when I'd found out that Meredith had been stabbed, hoping to prove that I knew the details of her death before an innocent person would have had the chance to.

It bothered me that as I answered him as fully as I could through an interpreter, Mignini would usually repeat the question. I was afraid I wasn't making myself clear. At first, Carlo, acting as a second interpreter, spoke in measured tones. He would interrupt and say, "What she is really saying is . . ." or "She's already answered that question!"

My lawyers listened intently to Mignini's wording, to his repetitions, to the interpreter's translation of his questions and my responses, and jumped up to object to suggestive phrasing and misinterpretations. They came prepared to protect me from what they'd warned me against: aggressive and insidious questioning by a prosecutor whose interest wasn't to hear me out but to get me to say something incriminating. Luciano and Carlo grew less measured as the interrogation dragged on.

After five and a half hours of standing and fielding questions, I was tired, but I thought everything was going okay. During the short breaks, Luciano would put his hand on my shoulder, and Carlo would say, "You're doing well."

Then the conversation turned to my November 6, middle-of-the-night interrogation and how I could have said something without meaning to. I explained how much pressure the police put me under and how confused I was by their claims that I'd met up with someone, that I'd been to the villa that night. Mignini became defensive. "I was there," he said, referring to the *questura* the night of my interrogation. "I heard you saying these things."

I said, *"You* were telling me these things. I was saying, 'I'm not sure. I'm confused.'"

This interrogation was becoming more and more like the one I'd meant to correct. It wasn't a do-over at all. Mignini would ask a question, and when I answered, he would reject my response and ask again. He was trying to intimidate me, spewing words at me.

Luciano and Carlo were leaning forward in their seats.

"Where did the name Patrick come up?" Mignini demanded.

"From my cell phone," I said. "Because I'd texted a message to Patrick. I wrote, 'See you later.'"

"What did you mean by your message?"

"In English, it means 'Goodbye. See you later, as in sometime.' It's not like making an appointment to see someone. And I wrote, *'Buona serata'*—'Have a good evening.' I had no plans to meet up with him."

"Why did you erase Patrick's message?"

"I sometimes erased the messages I received. I didn't have enough memory in my cell phone to keep them."

"Why did you say you didn't remember writing that message?"

"Because I didn't remember."

"Why did you name Patrick?"

"The police insisted I'd met the person I had sent the text message to."

"No. Why did you name Patrick?"

"The police had been asking me about Patrick."

"No! Why did you name Patrick?"

"The police insisted it was Patrick."

He was more and more aggressive about it. "Why Patrick?"

"Because of my message."

"That doesn't explain why Patrick."

"Yes, it does."

"Why did you say Patrick killed her?"

"Because I was confused. Because I was under pressure."

"NO!" he insisted. "Why did you say Patrick?"

I was more frustrated than I'd ever been. "Because I thought it could have been him!" I shouted, starting to cry.

I meant that I'd imagined Patrick's face and so I had really, momentarily, thought it was him.

Mignini jumped up, bellowing, "Aha!"

I was sobbing out of frustration, anger.

My lawyers were on their feet. "This interrogation is over!" Luciano shouted, swiping his arm at the air.

Carlo and Luciano sat me down and huddled around me, saying, "It's okay, Amanda, it's okay. You did a good job, and we'll talk about it the next time we come."

Then a guard walked me out. I was sobbing hysterically. I had done my best to explain everything, and I had failed completely.

As he left, Mignini apparently told waiting reporters that I hadn't explained anything or said anything new. All I did, he said, was cry.

That day changed everything for me. I understood that the prosecution's goal was not about trying to find out who had killed Meredith. I was left with the horrible certainty that I'd made a mistake and there was nothing I could do to fix it. There was nothing I could do that would make any difference to the prosecutor. In Mignini's hands, everything was distorted and bent to seem like more evidence of my guilt, and I was devastated.

Back in my cell, the Italian news channel was replaying a scene from the previous weekend, of Meredith's family, dressed in black, walking into her funeral service in England. I knew about the funeral from Don Saulo, and my spirit had been with Meredith all that day. As I watched her heartbroken family, I could only think, *With all I'm going through, I'm the lucky one.*

Chapter 21

January–May 2008

Clutching a garbage bag stuffed with my clothes and books, I stood at the gate of my third cell in nine weeks. The *agente* cranked the key in the lock and pulled. "What do you think this is?" she sneered. "A hotel?"

"No," I said, knowing that she saw my relocation requests as diva behavior.

I'd asked for the changes for solid reasons. My first cellmate, Gufa, had been erratic and difficult to live with. My next cellmates were three middle-aged gossips who criticized my cooking and cleaning. They called me a snob because I liked to read and write. "What good are your studies now, when you'll be spending the rest of your life in prison?" one asked.

They gave me a nickname: *Principessa sul Pisello*—the "Princess on the Pea." The reference to the fairy-tale title was a two-sided jab: *pisello* is a colloquialism for "penis," a reference to my supposed sexual depravity.

Now I was moving in with Cera. Young, with the tall, lean looks of a model, she worked as a *portavito*, delivering meals from a rolling cart. She was also in my weekly guitar class, another prison "rehabilitation" activity like movie time. But I was still

241

secluded from the main prison population—a special status to protect young, first-time suspects. The downside was that it prevented me from participating in group activities or talking to anyone but my cellmates. Thankfully, Don Saulo convinced prison officials to let me attend the guitar lessons, just as he had weekly Mass.

One Wednesday, as Cera and I walked back to our cells from our lesson, I asked, "Would you be willing to let me live with you? We're around the same age and we both study. I could help you with your English."

She waited a few beats before saying, "Sure. I'll write a request tonight."

Cera had managed to make her cell homey, clean, and organized. There were bright colored sheets on the beds, postcards taped to the walls, and a colorful curtain tied to the bars at the window. We had a heart-to-heart talk while I unpacked. She was sitting cross-legged on the bed closest to the window. "I should probably tell you right off, I'm bisexual," she said.

"That's cool," I replied. "I'm not, but I'm definitely live-and-let-live."

"You're not my type, anyway," she said. "I thought you might be gay when you asked to live with me, but I decided you weren't." She hesitated. "You know, your former cellmates said you're spoiled."

Wow. Why hadn't I realized they would trash me behind my back? They gossiped about everyone else. Cera read my disappointment. "They're fake. Almost everyone in prison is fake. You'll see."

"But it sounds like *you* have friends, that you have fun with people."

"What made you decide that?"

"I hear laughter coming from *socialità*." I wasn't allowed to go to the evening social time.

She rolled her eyes. "That's all bullshit. It's lighthearted, but everyone's fake."

How could everyone be fake? People are people.

"Prison is bad. You'll see." She leaned toward me. "Wait until you've been here awhile." She laid out the facts: "Prisoners and guards are in different worlds. The guards are the enemy. They're only here to judge us."

"They don't seem so bad," I said.

Cera scoffed. "You don't know what they say about you when you're outside—'Who does Kuh-nox think she is? She's saving worms from the rain but killing people.' Even Lupa says you're guilty."

I knew the prosecution didn't believe me, but I'd assumed the people I interacted with every day would see me for who I was and not imagine the worst. As soon as Cera said this, it seemed obvious —of course the guards would assume I was a murderer. Everyone did.

"The way to get along here is to appease the guards," Cera said. "Instill confidence in other prisoners. But mind your own business. And don't trust anyone."

I changed the subject. "Do you mind if I ask how old you are? And how long you've been in prison?"

She looked at me with the exaggerated patience of an adult speaking to a child. Until then, I didn't know that prisoners consider personal questions off-limits.

"I'm twenty-three," she said. "I've been in prison almost six years—of a twenty-five-year sentence. They say I murdered my boyfriend."

Oh my God. Hearing that made my heart hurt.

She continued. "I know how you feel, being the center of attention right now. Don't worry. They'll forget about you once the next sensational crime comes along."

As much as I wanted to be out of the limelight, the word *forget* terrified me, and hearing "twenty-five years" made my stomach lurch. I wanted to cry for her—and for myself. "Was the media tough on you?" I asked.

She flashed me another condescending look. "Journalists fixate on something and turn you into a symbol of evil," she said. "They say you have 'an angel face but a demon's soul.' Did you hear about Alberto Stasi?" He was accused of killing his girlfriend in August 2007. "Remember how the media reported that he has 'eyes of ice,' because they're blue. It was ice for me, too. They made me sound like a psychopath, because I like to chew on ice."

How am I still this naïve?

"Maybe we shouldn't talk about it," she said. "Save yourself the indignity."

Cera was right. When she talked she seemed angry and bitter. I didn't want to go there with her.

At twenty, I still had a childlike view of people. I looked for the saving graces in everyone. I thought people were naturally empathetic, that they felt ashamed and guilty when they mistreated someone else. That faith in humanity was being picked away, but I held to the belief that people were basically good. And that good people would believe me and set me free.

Part of the growing up I did in prison was learning that people are complicated, and that some will do something wrong to achieve what they think is right. Since my second interrogation with Mignini, I knew the prosecution was intent on undermining my alibi. Over the coming weeks and months, I would learn just how far they would go to try to prove me guilty.

In early January, Raffaele's father went on a popular Italian news program to convince viewers that his son had had nothing to do with Meredith's murder. "The bloody shoeprints in the villa were made by Rudy Guede," he said. The pattern of eleven concentric circles on the sole of Guede's Nike Outbreak 2s matched the prints on the floor. Dr. Sollecito produced a duplicate pair of the Nikes so TV viewers could see. A corresponding shoebox was found in Guede's apartment, he added.

The prints couldn't have been made by Raffaele's newer Nike Air Force 1s, he said. "They had just seven concentric circles." By show's end he had removed the possibility that Raffaele had been at the murder scene and put another strike against Guede. Raffaele's family must have felt euphoric.

But their elation didn't last twenty-four hours. The next morning, the prosecution announced new "evidence." The killer had slashed Meredith's bra off her body, slicing off a small strip of fabric that included part of the clasp. Raffaele's DNA was on the clasp.

"There's no way!" I said loudly. "It's impossible."

"I'm sure the police timed their announcement about the bra clasp to win the public back to their side after the show," Luciano said. "It's not a coincidence. Raffaele's lawyers made a terrible mistake going through the media instead of bringing their findings directly to the court."

I knew this "evidence" could hurt us. I also knew that Raffaele had as much chance of coming into contact with Meredith's bra as Meredith had meeting up with a knife from Raffaele's apartment. Neither could be true, but the prosecution would use both these findings to tie us to the crime.

"Raffaele's DNA must have been transferred to the clasp somehow," Carlo said. "Did you ever wear Meredith's clothes or share a load of laundry?"

"I borrowed tights and a shirt but never her bra," I answered. "And we washed our clothes separately. But we did dry them side by side on the same rack. Do you think that could be it?"

It turned out there was another possible explanation.

On December 18, forty-six days after the Polizia Scientifica first swept the villa for evidence, the Rome-based forensic police returned to No. 7, Via della Pergola.

Luciano and one of Raffaele's lawyers watched a live feed of the search from a van parked outside the villa. The investigators were dressed in white suits, shoe coverings, and gloves to protect the crime scene from contamination—but it was too late for that. The Squadra Mobile, or "Flying Squad," had already ransacked the house, tramping from room to room. While looking for Meredith's credit cards, keys, and other nonforensic clues, they'd dragged Meredith's mattress into the kitchen. Her unhinged armoire doors were on the floor. Her clothes were in heaps. The forensic team found the bra clasp under a rolled-up carpet, lying beneath a sock.

I wasn't implicated by the clasp, but I knew that the prosecution would never believe that Raffaele had acted without me. They'd say I gave him access to the villa. I was the reason he'd met Meredith. We were each other's alibis. If they could show that Raffaele was directly connected to the crime, I would, at the very least, be charged as his accomplice.

The bra clasp wasn't the only incriminating news the prosecution leaked to the press that day. "CSI Technique Leads Italian Police to Bloody Footprint in Foxy Knoxy's Bedroom," the London *Daily Mail* wrote. The article quoted Edgardo Giobbi, an investigator for the police, who said, "This is a crucial discovery and very important."

Luciano told me the low points. "They say your feet were 'dripping with blood'—that you tracked blood while you were trying to clean it up."

The forensic team used luminol, a chemical that glows blue when sprayed on even trace amounts of hemoglobin. It revealed two footprints in the hallway outside the bathroom and one in my bedroom.

"How can they say I had Meredith's blood on the bottoms of my feet?" I asked.

"Please don't worry, Amanda," Carlo said, giving me a sympathetic look. "I'm sure it's not as simple and straightforward as the media are portraying it. We've already spoken with our experts, and they say that you might have stepped on the blood splotch on the bathmat and tracked it down the hall. That could do it. And it's not just blood that shows up in luminol. It reacts the same to household cleaners, soil, juice, and rust from the faucet—anything that contains iron or peroxides. To know for sure what they're looking at, forensic scientists have to test separately with another chemical"—tetramethylbenzidine (TMB)—"that's sensitive only to human blood."

"Well, did they?" I asked anxiously.

"It's frustrating, but we'll have to wait until the investigation phase ends so we can see how the Polizia Scientifica reached their conclusions," Carlo answered.

Perhaps it was better that we didn't know then it would be

twenty-two nerve-rattling months before we found out how the forensic scientists had made this misleading call.

This new claim was another barricade separating me from my real life—one more accusation on a growing list. Too many impossible things were being served up as "truth"—Meredith's DNA on Raffaele's kitchen knife, Raffaele's DNA on Meredith's bra clasp, and now Meredith's blood on the soles of my feet.

It was crazy enough to be told that "investigative instinct" had convinced the police I was involved in Meredith's murder—that I was dangerous and evil. Now forensic science—the supposedly foolproof tests I was counting on to clear me—was turning up findings I knew were wrong. I, like most people who get their information from TV crime shows, was unaware that forensic evidence has to be interpreted, that human error and bias can, and do, upend results.

"I don't get it," I told my dad at his next visit. "How can this be happening? Raffaele and I weren't there, so how can there be *any* evidence pointing to us?"

I felt so weighed down, so helpless and sad, that all I could do was cry while Dad held me. "Are the police just really bad at their jobs?" I wailed. "They're getting further and further away from the truth. How can the investigators make three incriminating errors in a row? What will they find next?"

But I tried not to think too far ahead. I'd already had to tell Mom good-bye. She's an elementary school teacher who had used all her vacation and sick days to be with me. My defense was costing far more than my parents had. She had to get back to work. Thank God Dad was there for me. I couldn't fathom how I would get through this without my parents.

Incensed by the stream of falsehoods, I concluded what my lawyers and my dad already knew: the police and the prosecutor couldn't afford to admit they'd made a mistake. They'd announced, "Case closed," at the press conference the day Raffaele and I were arrested. They would stick to their story at all costs.

I always liked seeing my lawyers, but now I had to brace myself for each visit. I didn't have to wait long before they brought more devastating news.

Less than a week later, investigators reported that they'd found my DNA mixed with Meredith's blood ringing the drain of the bidet in our shared bathroom. The implication was that I'd rinsed my hands and feet in the bidet after slashing her throat. They said that my skin cells had shown up—not Raffaele's or Rudy Guede's—because I was the last person to wash up in that bathroom.

The other update that day was something my lawyers had learned about when an Italian reporter held up his cell phone to show Luciano a series of photos in that day's *Daily Mail.* "Chilling Pictures of Meredith Murder Scene Reveal Apartment Bloodbath Horror," read the headline.

When I'd come home from Raffaele's on November 2 there were two dots of blood in the sink and a tiny smear on the faucet. In one of the *Daily Mail* photos, the bathroom where I'd showered appeared to be drenched in blood. Police released the photo with no explanation. They didn't say that the room had been sprayed with phenolphthalein, a chemical that, like luminol, is used as a first screen to detect the presence of blood. Also left out was the fact that phenolphthalein immediately turns certain bases and acids, including hemoglobin, a pinkish red. Thirty seconds after that, everything touched by phenolphthalein—every wall, every floor tile, every fixture, every towel—turns that lurid shade. I could only conclude that the police had distributed the pictures of the bathroom knowing that most people would never have heard of the chemical and would, naturally, believe the red was blood.

The pictures of the chemical-stained bathroom did what, I have to assume, the police wanted. The public reaction proved that a picture—especially a "bloody" picture from a crime scene—is worth a hundred thousand words. At least. I knew what people were thinking. *Who but a knife-wielding killer would take a shower in a "blood-streaked" bathroom? Who but a liar would say there had been only a few flecks of blood? The answer? Foxy Knoxy.*

The bathroom photos were released along with pictures of Meredith's room, both before and after her body had been taken away. There were photos of the bloody shoeprint that was still being attributed to Raffaele, even after his family had proven it couldn't be his. One photo offered an almost complete view of the room from the doorway; another showed Meredith's naked foot sticking out from under her comforter. Close-ups showed the tremendous amount of blood Meredith lost, choked on, and died in. Seeing these shots made me weep. *She must have been so scared.*

The public doesn't usually have the right to see the prosecution's documentation until the defense does. But the photos were out, and there was no way to dampen the effect. It struck Luciano as another attempt by the prosecution to win public favor.

My lawyers complained to the judges that the prosecution was using the media to our disadvantage, but the judge said that whatever was reported in the press wouldn't be held against us. The flow of information between the prosecution and the media was an accepted but unacknowledged fact.

Playing elite soccer as a teenager had taught me that to beat the opposing side, I'd need maximum endurance, persever- ance, and tenacity. I started thinking of myself as part of a team led by my lawyers; I had to help them succeed. Drawing on the little reserve I had left, I willed myself through the emotional pain. When I was seventeen, I played for a month on a broken foot before admitting it to my coach. I felt like that now: deter- mined but vulnerable.

The denial, fear, and bafflement I felt in the beginning of this nightmare had turned into quiet indignation and defiance. I fi- nally accepted that I was my only friend inside Capanne. I clung to my dad at every visit. The rest of the time, I used the only coping tool I knew: I retreated into my own head.

The natural reaction to having no control over your own life is to grab on to ways to feel that you do. In prison the only thing you're in charge of is your body. You can overexercise. You can hurt it. You can overeat. You can starve. You can decide what goes in and what stays out. I refused to let antidepressants or sedatives cross my lips. And I went silent.

After nearly five months at Capanne, the only people I talked to consistently were my family on visiting days and Don Saulo, when I saw him (my only stress-free moments in prison). Otherwise I answered questions; I didn't ask them. I didn't comment. Memo- ries of my real life at home were my sanctuary. I didn't want to mix it up with this miserable faux life I was living behind bars.

Cera's sense of control came from cleaning. When I moved in I liked that her cell was spotless. I didn't understand that it was her obsession, until she demanded that I dry off the walls of the shower before I dried myself; place the shampoo and lotion bottles in a perfect line on the counter, equally spaced apart; tuck in my bedsheets with military precision; arrange the apples in the fruit bowl stem up; and avoid using the kitchen sink.

I tried hard to get along with Cera. I helped her with her schoolwork and either cleaned alongside her or stayed out of the way. My job, after she was done mopping and drying the floor, was to take a *panno spugna*—a spongelike cloth—and clean the baseboards on my hands and knees. I complained bitterly to Mom about these things when she came to Italy over her spring break.

One morning, when I was walking into the bathroom to put something away, I bumped into Cera, and she kissed me on the lips. I just stood there staring at her, too surprised to know what to say. "Your face is telling me that was not okay," she said quickly. "I'm really sorry."

She never made physical advances after that, but she did once ask if I was curious what it was like to have sex with a woman, like her. My stock answer—an emphatic no—made her feel bad.

I told Mom about that, too.

"Amanda, we need to talk," Cera said one day. She was leaning in the kitchen doorway, watching me stare at the wall, her arms crossed over her thin torso. "Look, I don't feel like we have a relationship. Why don't you talk to me?"

"I honestly don't have anything to say," I said. "Everything I think about is really personal," I stammered, my eyes starting to tear up.

I no longer trusted the authorities. They were against me. I was continually under surveillance. I read. I practiced Italian. I spent most of my time writing letters to the people I desperately missed—my mom, my dad, Madison, Brett, DJ, Oma, my sisters. It was the only way I felt connected to anything outside prison.

How could I explain this to Cera?

"When I look at you I see myself four years ago," she said. "I don't care if you're guilty or not, but I worry about whether you suffer. I don't want you to make the same mistakes I did. In the years I've spent in prison, I've screamed, fought, starved, and cut myself, and no one cared or made the effort to help me. Please come out of your shell before it destroys you. If you're always hiding inside yourself, you won't ever be able find your way back out."

———

My only hope and constant thought during that winter and spring was that the judge might allow me to live with my family in an apartment, under house arrest. My first plea had been rejected, but my lawyers had another hearing scheduled for April 1. Even though Carlo and Luciano weren't confident about the outcome, I was sure it would happen. I was counting the days.

Less than a week before the hearing, I heard on TV that Mignini had interrogated Rudy Guede again. I listened to the newscast, hoping Guede would tell the truth.

My heart started pounding as I listened. "Amanda and Raffaele were at the house that night," Guede reportedly said. "I saw them. When I came out of the bathroom, I saw a male figure. I put my hand on his shoulder, and he had a knife in his hand. I also heard Amanda Knox. She was at the door; I saw her there. The two girls hated each other. It was a fight over money that sparked it off. Meredith accused Amanda of stealing three hundred euros from her drawer."

"That's a total lie!" I burst out. I'd never felt so much hatred for another person as I did toward Guede in that moment.

Cera looked over at me with a pitying glance. "Now you're really screwed," she said. "Once defendants start blaming each other, it's all over—for him and for you. That's what the prosecution wants. That's how they make it impossible for you to defend yourself."

Luciano and Carlo came to see me the next day. They reassured me that no one, not even the prosecution, believed Guede. "He ran away, he's a liar, a thief, a rapist, a murderer," Carlo said. "No one could ever consider him a reliable witness, because he has everything to gain from blaming you. The prosecution is making a big deal about it because it incriminates you."

"Please, Amanda," Luciano said. "This is not what you need to worry about. You need to stay strong."

Still, I couldn't be consoled. With Guede's testimony against me, there was absolutely no chance a judge would free me from prison.

In early April, Carlo came to Capanne. His face gave away his worry. "Amanda," he said, "the prosecution now says there's evidence of a cleanup. They contend that's why there's no evidence that you and Raffaele were in Meredith's bedroom—that you scrubbed the crime scene of your traces."

"That's the most ludicrous reasoning I've ever heard!" I screeched.

"Amanda, the investigators are in a conundrum," Carlo said. "They found so much of Guede's DNA in Meredith's room and on and inside her body. But the only forensic evidence they have of you is outside her bedroom. Raffaele's DNA evidence is only on the bra hook. If you and Raffaele participated in the murder, as the prosecution believes, your DNA should be as easy to find as Guede's."

"But Carlo, no evidence doesn't mean we cleaned up. It means we weren't there!"

"I know," Carlo said, sighing. "But they've already decided that you and Raffaele faked a break-in to nail Guede. I know it doesn't make sense. They're just adding another link to the story. It's the only way the prosecution can involve you and Raffaele when the evidence points to a break-in and murder by Guede."

Judge Matteini sent me her decision about house arrest on May 16: "Denied." By then the prosecution had stacked so much against me that Guede's testimony hadn't even figured in her decision. Even though I hadn't left the country before my arrest, the judge was certain that Mom would have helped me leave when she was to have arrived in Perugia on November 6. That, she said, is why the police planned to arrest me *before* Mom could get to me. It turned out that they'd gotten her itinerary the same time I did—by bugging my phone.

Before concluding, the judge criticized me for not showing remorse for Meredith's death.

When Carlo and Luciano came to tell me my request for house arrest had been denied, my mind rolled back to the *questura* on the morning of November 6. After my interrogation had ended, I was distraught and whimpering, sitting in the empty office with the lead interrogator, Rita Ficarra. My cell phone started ringing, vibrating loudly against the desktop, and I'd begged Ficarra to let me answer it. I was sure that it was my mom, and I knew she'd be undone with worry if I didn't pick up.

This new setback conjured up all the desperation, the nauseating helplessness, I'd felt that morning. I could hardly breathe thinking about it. I remembered how relieved I'd been that my mom was flying over, how much I needed her. As soon as she said she was coming to Italy, I realized I'd been stubbornly, stupidly insistent that I could help the police find Meredith's killer on my own.

I'd been tricked.

I understood that this regret went beyond me. My mom was eating herself up with guilt for not having come sooner. When I saw her over her spring break, she'd lost twenty pounds. She wept at every visit.

After the judge's decision, everything seemed darker. I talked to Don Saulo a lot about how claustrophobic I felt with the possibility of house arrest off the table. I couldn't concentrate on reading, Italian grammar, or even on writing letters home, for all the anger, disappointment, and sadness I felt.

Cera started trying to prepare me for the chance of another fifteen years in prison. "I think you should say you're guilty," she advised me one day, "because it will take years off your sentence."

"I will not lie!" I yelled, spitting out one word at a time. "I'm not scared of Guede or the prosecutor! I'm ready to fight! I don't know anything about this murder, and I will go free!"

Luciano and Carlo tried to steel me for what they knew would eventually happen.

"You have to be ready to take this case to trial," Carlo said one day in May, his finger poised over the mouse pad of his laptop. "The prosecution is going to say things about you. You're going to see and hear all the horrible details of Meredith's murder. It's going to be tough on you, Amanda."

With that, he turned his computer around for me to see. He scrolled down. Meredith's face, tilted upward, showed up yellow and wide-eyed on the screen. A grotesque, dark, gaping gash seemed to burst from her neck.

Ah! I gasped and turned away. I felt as if I were choking.

Carlo half-rose and said, "Amanda, calm down."

I struggled for breath that came in painful hiccups. "I can't!"

"We should call it a day," he said, standing. He knocked on the door. *"Assistente!"*

I could not stop wailing.

Carlo helped me out of my chair. He held his hand gently against my back when the *agente* opened the door.

"It's a rough day," he explained.

The *agente* grasped my shoulders firmly and steered me around the corner, almost into the *ispettore*—"supervisor"—who was walking down the hall.

"What's wrong with you? What happened?" she asked.

"I saw Meredith's autopsy photo."

"What?"

"Meredith's autopsy photo," I mumbled.

The *ispettore* looked at me bewildered. "But you've already seen her dead!"

I wanted to break away from the *agente*'s grip. The *ispettore* thought I had killed Meredith. Everyone thought I'd killed Meredith.

I wanted to go back to my cell, to be by myself. I wanted everyone to stop looking at me. I wanted to breathe. I couldn't get Meredith's face out of my mind—the complete absence of expression, the grayish yellow tone of her skin, the dark and vivid red of the wound. I couldn't reconcile the Meredith I knew with the image I'd just seen.

Instead of walking me to my cell, the *agente* led me into the infirmary and directed me to sit down in front of the doctor on duty.

"What happened?" he asked, leaning forward.

"Meredith's autopsy photos," I said, my hysteria having dwindled to a sniffle. "I just saw them for the first time."

"I can prescribe a sedative for you."

"No. Please, I just want to go back to my cell."

He paused a moment, then met the *agente*'s eyes. "As you wish," he said.

Chapter 22

June–September 2008

E verything—and nothing—changed the morning in late June when I was called downstairs to sign yet another document. The guard barely raised his eyes while pulling out the paperwork and pointing to the line awaiting my signature. When I finished, he handed me the last copy from the stapled pile. I recognized Mignini's illegible scrawl and Judge Matteini's loopy cursive that always made the *M* look like a *W*. *Watteini*.

It was only after I went back upstairs and sat down on my bed that I read:

—NOTICE OF THE CONCLUSION OF THE PRELIMINARY INVESTIGATIONS—

The Prosecutors Dr. Giuliano Mignini and Dr. Manuela Comodi;

considering the documents in the proceedings indicated in the epigraph registered on 6/11/2007 in regard to:

Amanda Knox

KNOX, Amanda Marie, born in Seattle (the state of Washington—USA) on 7/9/1987 . . . presently detained in the Casa Circondariale Capanne of Perugia;

SOLLECITO, Raffaele, born in Bari on 3/26/1984 . . . presently detained in the Casa Circondariale of Terni;

GUEDE, Rudy Hermann, born in Agou (the Ivory Coast) on 12/25/1986 . . . presently detained in the Casa Circondariale of Perugia;

persons subject to the preliminary investigations all, for having, in collaboration, murdered Kercher Meredith by strangulation and a profound lesion by a pointed, cutting weapon . . . and taking advantage of the late hour and the isolated position of the apartment . . . and having committed the act for trivial reasons while Guede, in collaboration with the others, committed rape; Knox and Sollecito, for having, in collaboration, carried out of Sollecito's apartment, without justifiable reason, a large, pointed, and cutting knife; Guede for having, in collaboration with Knox and Sollecito, forced Kercher Meredith to suffer sexual acts, with manual or genital penetration, by means of threat and violence; all because, in collaboration, for having procured for themselves an unjust profit by having taken possession of a sum of 300 euro, two credit cards, and two cell phones, all belonging to the same Meredith; Sollecito and Knox, for having, in collaboration, simulated an attempted robbery with the break-in in the bedroom of Romanelli Filomena, breaking the window with a rock found around the house and left in the room, near the window, all in order to assure themselves impunity for the crimes of murder and rape, attempting to attribute the responsibility to strangers having entered the apartment.

Events having occurred the night between 1 and 2 November 2007.

Knox, for having, with further acts executed in the same criminal design, knowing him to be innocent, in declarations made to the Police Flying Squad of Perugia on 6 November 2007, falsely blamed Diya Lumumba, called "Patrick," of the murder of Kercher Meredith, in order to assure herself impunity for everyone and in particular Guede Rudy Hermann, also of color like Lumumba, in Perugia the night between 5 and 6 November 2007.

NOTIFY
the persons subject to the preliminary investigations: that the preliminary investigations are concluded.

Oh my God. I've been formally charged with murder.

I wanted to scream, "This is not who I am! You've made a huge mistake! You've got me all wrong!"

I was now fluent enough in Italian to see how ludicrous the charges were. Along with murder, I was charged with illegally carrying around Raffaele's kitchen knife. It was galling. Real crimes had been committed against Meredith; the police owed her a real investigation. Instead, they were spinning stories to avoid admitting they'd arrested the wrong people.

I shouldn't have been thrown when I received these formal charges. For nearly eight months, I'd been jailed as a suspect. I'd been expecting my indictment to be sent down since the awful day when Carlo had made me face up to the gruesome autopsy photos.

But a tiny part of me had held out hope that when Mignini spread all the evidence before him, he would see that his theory didn't hold up.

Luciano and Carlo came to see me soon after.

"Now's our chance to stand up and fight," Luciano said, punching the air. "This is what we've been waiting for."

Finally we could combat all the misinformation leaked to the media. We could explain that the knife had never left the kitchen, the striped sweater had never gone missing, the receipts weren't for bleach, the underwear I bought wasn't sexy. We could describe how the prosecution had come up with the bloody footprints. We'd explain why Meredith's blood had mixed with my DNA in our shared bathroom, how my blood got on the faucet, and correct the notion that the crime was a sex game gone wrong. We could object to the prosecutor painting me as a whore and a murderer. My lawyers would finally get to see the prosecution's documents. *No more surprises.*

"Our forensic experts are already reviewing the files to prepare for the pretrial in September," Carlo added. "Now that the investigation's over, we'll have a different presiding judge. We hope whoever it is will have a better sense of logic than Claudia Matteini."

"You have to be kidding me! We have to wait all summer?" I moaned.

That's when I found out that the Italian courts shut down almost completely for the last half of July and all of August.

I spent that afternoon jogging alone: round and round in small, dizzying circles in the courtyard outside the chapel. I'd long ago figured it took about eighty laps to make a mile. Suddenly, Argirò opened the door. "Kuh-nox," he called, waving me inside.

Odd. Prison is all about routine, and this had never happened before.

"What's going on?" I asked, confused.

"We're taking you off your restricted status."

Just like that. While I was being investigated, I was under judge's orders to be kept separate for my own safety. But now, as an accused criminal, I passed from the judge's responsibility to the prison's.

Up to that moment, I didn't believe this would ever happen. Only a few days had passed since I'd been moved out of Cera's cell due to mutually agreed-upon incompatibility—I wasn't fastidious enough for her, and to me, she was intolerably controlling—and switched to a cell with two big-bosomed, middle-aged sisters named Pica and Falda, who defined themselves with the politically incorrect word *zingare*—the feminine for "gypsies." They were kind and uneducated—neither had learned to tell time, and when I tried to explain that Seattle was on the other side of the globe, they didn't know what I was talking about. Finally, I realized they didn't know the earth was round.

Prison officials had always claimed I was kept separate—I had cellmates but, with the exception of a few prescribed events, couldn't interact with the broad population—because other inmates would probably beat me. Now, with only the mildest caution—"Be careful of the other girls!"—Argirò opened a second door. Instead of having *passeggio* by myself, I was in the company of fifteen sweaty women.

As soon as I walked outside, the gaggle of prisoners started hooting and hollering, "She's out! She's with us! Way to go!"

I was in a concrete-walled area about a third the size of a football field. The ground was covered with hard, orangey-purplish-red rubber. It was the angriest red I'd ever seen—and bare except for a few white plastic benches and dozens of cigarette butts. I didn't care. This was the most open space I'd seen since coming to prison. I took off in a sprint, making wide loops, skipping, and whooping, "I'm out! I'm out!" My fellow inmates stared, probably thinking I was just as incomprehensible as the media made me out to be.

I introduced myself to women I'd seen around Capanne—at movie time or Mass or guitar class—but hadn't been able to meet. I'd had only my cellmates for company before, and those relationships were ultimately frustrating and upsetting.

At 3 P.M., when *passeggio* ended, we lined up to be patted down by an *agente*. A girl I didn't know came up to me. "I'm Wilma," she said. "Will you buy me two packs of cigarettes?"

"I guess," I mumbled. Caught off-guard, I didn't know what to say.

I amended. "I'll buy you one pack."

That night, I went to my first *socialità* full of pent-up energy I didn't even know I had. Being thrust in with all these new people—talking and playing Foosball and cards—reminded me of my freshman year in high school. *All I have to do is find my clique and get along.*

My excitement didn't last long. A couple of women came up and started heckling me. "Why are you buying cigarettes for Wilma?" they demanded. "She doesn't deserve anyone's help."

That started a chorus of grumbling: "Fricking *infame*."

Infame—"infamous person" or, in prison, "snitch"—was the worst label you could have there. At best you'd be ostracized. At worst you'd be abused by other prisoners.

Wilma, it turned out, was an outcast in this small circle of prisoners. I didn't know her story, but I felt bad for her. Just as in high school, when I hung out with the less popular crowd, I instinctively sided with her. I spent hours listening to her mope about how sad and confused she was. One day she said, "Amanda, can you explain why everyone hates me?"

By then I'd heard enough of the gossip to figure it out. "It's because you talk about people behind their backs and tell on prisoners to the guards," I explained. "Maybe you can change your behavior and people will start liking you."

Just like high school.

I didn't expect her outburst at *socialità* that evening. Wilma screamed, "All you people talk badly about me."

Another prisoner came up to me and demanded, "Why did you tell Wilma everyone hates her?"

I said, "Well, it's true, isn't it?"

"Yes, but you're not supposed to talk to her! Why did you side with her?"

Wilma's behavior wasn't that different from that of other prisoners—most were manipulative and liked to stir up drama—but she wasn't smart enough to recognize this and to fake loyalty to the other women. People were able to see through her actions.

R affaele was charged the same day I was. But I was so consumed by figuring out how to navigate this new larger prison world, I hadn't given him much thought outside of the facts of the case. Within days of our indictment an envelope with his name printed on the back arrived for me. I had never seen his handwriting before, and at first I suspected it was a nasty joke.

As soon as I read the letter, I realized it was real. I was shocked that he was writing me. I'd felt betrayed by the months of silence and by his comments in the press distancing himself from me. And of course there was the issue of his previous claim that I had left his apartment the night of the murder and asked him to lie for me.

He wrote that he'd been aching to contact me, and that it was his lawyers and family who hadn't permitted him to get in touch. He said everyone had been afraid when we were first arrested, but that now he realized it had been a mistake to abandon me and wrong to submit to police pressure and acquiesce to their theory. "I'm sorry," he said. "I still care about you. I still think about you all the time."

I understood. My lawyers had given me the same strict orders.

I felt completely reassured by his letter. It wasn't lovey-dovey, and that suited me fine. I no longer thought of us as a couple. Now we were linked by our innocence. It was a relief to know we were in this fight together. It was only much later that I learned how his interrogation had been as devastating as mine.

I wrote him back the next morning. I was explicit about not wanting a romantic relationship anymore but added that I wanted the best for him and hoped he was okay. I knew I shouldn't write about the case, so I only said I was optimistic that our lawyers would prove the prosecution wrong.

As soothing as my correspondence with Raffaele was, I got another letter that summer that undid me, making me realize again how much my situation was affecting my family. My youngest sister, Delaney, who was nine, wrote, "Dear Amanda, I was at the pool a few days ago with Mom, Dad, Ashley, and Deanna. A boy came up and asked if they were my sisters. I said, 'Yes, but there's Amanda, too.'

"The boy said, 'That sister doesn't count.'

"It made me so sad. What should I say when someone's mean about you?"

Delaney's letter had taken two weeks to reach me. My reply wouldn't get to her for another two. I was as low as I'd been since my first days in prison.

Still, it meant a lot that she'd asked my advice. As the oldest, it meant the world to me that my sisters came to me when they were upset. I was afraid that connection had been lost. I was terrified my family would stop being honest with me for fear that it would somehow wound me.

———

Besides the prison *vice-comandante*, Argirò, men were rarely allowed in the women's ward. One exception was the workers who came on Fridays to fix plumbing and electrical problems. When I was living with Cera, the guard in charge, Luigi, told her he thought I was cute. He often stopped to chat. Once, he sat on my bed and waved in his workers to have their cigarette break in our cell.

On July 4 the shower in my new cell was clogged. I didn't know how to say "drain" in Italian, so I said, "The hole in the shower won't let the water down."

"What hole?" Luigi asked.

I was alone—Pica and Falda were at their prison jobs—and Luigi followed me into the bathroom. As soon as the door closed, he grabbed me around the waist and pulled me to him, leaning forward as if to kiss me. I ducked my head and went stiff, as though a steel rod had been jammed down my spine. Somehow I managed to wriggle out of his grasp. I stumbled out of the bathroom, sat on my bed, and pulled my knees to my chest, shaking.

He didn't look at me as he came out of the bathroom. He just mumbled that his guys would fix the shower and left.

I knew I couldn't tell anyone. Luigi could turn this incident against me. What if he called me a liar? Or said I'd come on to him, that I was obsessed with sex—as the prosecution was saying? No one would believe me.

I went to the window and cried—not out of sadness, but from a place of deep, black anger. It was one thing to have people saying things about me on TV and another to be overpowered.

The unwanted attempt to kiss me happened five days before my birthday. *So this is the gift the prison has to offer me,* I thought cynically. *A reminder of my helplessness.*

My mom and Deanna came to visit me on July 9, the day I turned twenty-one. They sang "Happy Birthday," but bringing in cake was not allowed. "Don't worry, Amanda," Mom said, putting her best spin on it. "We're not celebrating anything until you get home. And I'm sure that will be soon."

I erupted into sobs when I hugged my mom and sister goodbye. Back in my cell, I'd barely pulled myself back together when I was called down to the ground floor again. Raffaele had sent me a huge bouquet of white lilies. The guards were shaking their heads and chuckling about it, as though they'd never seen anything so absurd. When I reached for the vase, the guard said, "Prisoners aren't allowed to have flowers."

I guess that was another hiding place for drugs.

Nothing eased my pain that day. Ever since I was a little girl, I'd always dreamed of being older, counting off the years until I'd be a teenager and then again to the day I'd finally be grown up. The previous year, when I turned twenty—not long before I left for Italy—everyone in my family had said, "Next year's birthday is going to be the real party."

Instead, here I was now, literally sweating out my twenty-first birthday in a sweltering Italian prison. I learned strategies from the other women for how to stay cool. We took frequent cold showers, wetting our hair over and over. We drenched our sheets and tied them to the bars of the windows in a vain attempt to cool any whisper of a hot, dry breeze that might pass through.

All this happened while Luciano and Carlo were preparing the defense for my pretrial. They didn't have everything they needed to break down the case completely—Meredith's DNA on the knife and my "bloody" footprints were going unanswered.

Two days before the pretrial started, we got news that was both heartening and unnerving. Police investigators revealed that they'd found an imprint of the murder weapon in blood on Meredith's bedsheets, making it clear the weapon wasn't in fact the knife with the six-and-a-half-inch blade the prosecution was claiming. The imprint was too short to have been made by Raffaele's kitchen knife.

I'd thought that being charged marked the true end of the investigation. Now I felt catapulted back to the spot I'd been for the past several months—back to wondering what the prosecution was going to spring on us next.

At the same time, we had evidence that Carlo called "the murderer's signature."

I reminded myself that we also had common sense on our side. There was no motive. I had no history of violence. I'd barely met Rudy Guede. Raffaele had not met him at all.

Luciano and Carlo came to see me the day before the pretrial started. "We'll be as strong and forceful as we can," Luciano promised.

Carlo, the pessimist, said, "Don't get your hopes up, Amanda. I'm not sure we'll win. There's been too much attention on your case, too much pressure on the Italian legal system to think that you won't be sent to trial."

Chapter 23

September 18–October 28, 2008

T he *agente* slid the cage door closed and turned the lock. Terrified and claustrophobic, I was alone inside a cramped, cold metal box barely large enough to sit in. I put my mouth up to the honeycomb panel and sucked in air. I heard the van's double doors slam and felt the vehicle lurch as we pulled away from Capanne. The four blue-uniformed guards could see the countryside; I could not. I knew we were twisting and turning our way to the courthouse in downtown Perugia for the pretrial that would decide if the prosecution had enough evidence against Rudy Guede, Raffaele, and me to send us to trial.

The excursion made me feel more trapped than I felt in jail. I was like a package on a FedEx truck—on board but untended. The guards' job was to deliver me. Nothing more.

I longed to look out the window. I'd been outside the prison only once since my arrest, and that was only because I needed to see an eye doctor. Being in prison where the vistas are all broken by bars had made my eyes nearsighted. During that earlier trip, I heard children playing, and I started sobbing. It made me think of my family and of all I'd lost.

The pretrial was a far more jarring experience.

As the van rolled down the ramp and into the courthouse's underground garage, one *agente* said, "The journalists are waiting for you, Kuh-nox."

"You're going to be a good girl so we don't have to handcuff you, right?" another guard said. I had always been so polite and docile that a guard had once said to me, "If all the inmates were like you, we wouldn't need prisons."

I'd thought to myself, *Because I shouldn't be a prisoner.*

Between leaving the van and entering the double doors of the courthouse, I had a few moments in the open air but not free rein. A guard held my arm, using it to steer me into the building and, from there, into the antechambers used to confine prisoners. Later, in the courtroom, one guard stood behind me. Another was stationed a few steps to the side.

Walking down the hall, I was sandwiched between two guards, with a third *agente* ahead of us. "Remember. Do not say anything to the press. Don't even look at them," one cautioned.

As we rounded a corner, cameras flashed. Media people were yelling questions in Italian and English—"Are you guilty?" "Are you innocent?" "Why did you do it?" "Did you do it?" "What happened to Meredith?" "What do you have to say?"

The journalists and photographers were barricaded behind a rope. I didn't notice their faces, only huge black camera lenses and blindingly bright explosions of light. I felt so self-conscious that I instinctively ducked to hide my face.

And then it was over. The courtroom—closed to anyone who wasn't involved in the case—was pin-drop quiet. My team's table was on the far right. The table for Raffaele's lawyers was next to mine. Rudy Guede was to sit with his lawyers behind Raffaele's table. But on the first day, I was the only one of the three defendants there.

Relieved that the room wasn't full of people, I sat down and waited for the judge. Then the double doors I'd come through opened again, and the Kercher family walked in.

My first thought wasn't *They think I'm a murderer.* It was *Meredith's parents? I finally get to meet them.*

I asked Carlo if it would be okay to say hello. He checked with their lawyer, Francesco Maresca, and came back saying, "Maresca said, 'Absolutely not.' Now is not the time."

Will there ever be a right time?

I knew I had to listen to my lawyers on this one, but I was still waiting for a moment when we could exchange glances so that I could convey my sympathy.

When I tried to catch their attention, they glared at me, and I felt as if I'd been slapped. I also felt completely humbled. Meredith's mom's expression was both hard and sorrowful.

I was devastated. I'd anticipated meeting them for a long time. I'd written and rewritten a sympathy letter in my head but had never managed to put it on paper. Now I felt stupid. How had I not anticipated their reaction? *Why are you so surprised? What do you think this has been about all along?* My grief for Meredith and my sadness for her family had kept me from thinking further. *Of course they hate you, Amanda. They believe you're guilty. Everyone has been telling them that for months.*

The first day of the pretrial was mostly procedural. Almost immediately Guede's lawyers requested an abbreviated trial. I had no idea the Italian justice system offered this option. Carlo later told me that it saves the government money. With an abbreviated trial, the judge's decision is based solely on evidence; no witnesses are called. The defendant benefits from this fast-track process because, if found guilty, he has his sentence cut by a third.

Guede's lawyers must have realized that he was better off in a separate trial, since the prosecution was intent on pinning the murder on us. The evidence gathered during the investigation pointed toward his guilt. His DNA was all over Meredith's room and her body, on her intimate clothing and her purse. He had left his handprint in her blood on her pillowcase. He had fled the country. The prosecution called Guede's story of how he "happened" to be at the villa and yet had not participated in the murder "absurd"—though they readily believed his claims against Raffaele and me. One of the big hopes for us was that with so much evidence against Guede, the prosecution would have to realize Raffaele and I hadn't been involved.

I felt the way about Guede that Meredith's family felt about me. As soon as I saw him, in a subsequent hearing, I thought angrily, *You! You killed Meredith!*

He didn't look like a murderer. He was wearing jeans and a sweater. It was almost impossible to imagine that he had cut Meredith's throat. But if he hadn't, his DNA wouldn't have been everywhere in Meredith's room. And he wouldn't have lied about Raffaele and me. The other thing I noticed: he wouldn't look at me.

I was relieved when Raffaele appeared on the second day of the pretrial. He smiled as soon as he saw me, as though he couldn't suppress it. After almost a year being apart from him, my second impression was the same as my first: he was honest and smart—and, even with shoulder-length hair, just as handsome as when I

Casa Circondariale Capanne di Perugia, where I was imprisoned for four years. *(Oli Scarff/Getty Images)*

My parents, Curt Knox and Edda Mellas, surrounded by reporters outside the prison in November 2007, days after my arrest. My parents were divorced but came to my aid together. *(AP Photo/Leonetti Medici)*

My family leaves Capanne after visiting me. *From left:* my sisters Ashley and Deanna, my mom and dad, and my sister Delaney.
(Oli Scarff/Getty Images)

Don Saulo Scarabattoli, the Catholic chaplain for Capanne's women's ward, and my dear friend.
(Courtesy of Don Saulo Scarabattoli)

Being escorted by guards to Perugia's courthouse during my pretrial in September 2008, after ten months in prison. *(AP Photo/ Pier Paolo Cito)*

Co-prosecutors Manuela Comodi and Giuliano Mignini. *(AP Photo/ Stefano Medici)*

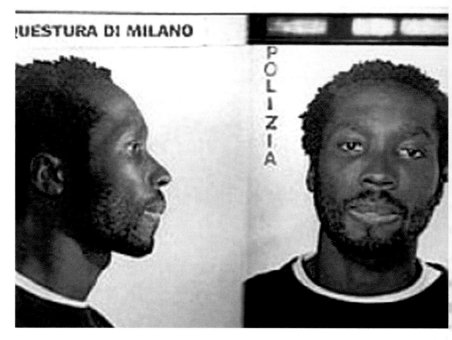

Rudy Guede's mug shot. He was convicted for his involvement in Meredith Kercher's murder in a fast-track trial in September 2008.
(Source: Perugia Police Department)

Raffaele's kitchen knife, which the prosecution alleged to be the murder weapon. Court-appointed experts cleared it in June 2011.
(Source: Perugia Police Department)

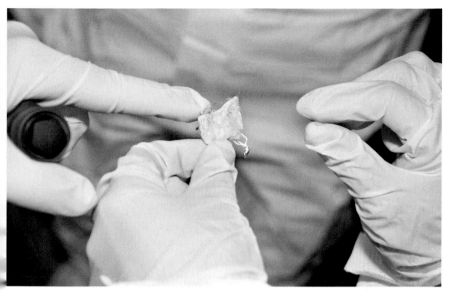

Wearing soiled gloves, members of the Polizia Scientifica hold a section of Meredith's bra left at the crime scene for six weeks after her body was found. Raffaele's DNA found on the dangling hook was the result of contamination. *(Source: Perugia Police Department)*

My father retrieving my things from No. 7, Via della Pergola, early eighteen months after Meredith was found murdered and the villa was sealed. © *Daniele La Monaca/ Reuters/Corbis)*

After a year and a half in prison, I took the witness stand during my trial, testifying in Italian, without an interpreter. *(Franco Origlia/Getty Images)*

My mother arriving in court with my lawyers, Luciano Ghirga *(left)* and Carlo Dalla Vedova *(right)*. *(Franco Origlia/Getty Images)*

Chatting with my
lawyers during a break
at the courthouse.
A guard is ever-present.
(Franco Origlia/Getty Images)

Arriving at the courthouse during closing arguments
days before my conviction in December 2009. Raffaele,
behind me, let his hair grow out in prison.
(Tiziana Fabi/AFP/Getty Images)

My mom, Deanna,
and my dad during
closing arguments in
December 2009.
My entire family came to
Perugia for the verdict.
(Giuseppe Bellini/Getty Images)

In the prison van on my
way back to Capanne just
after my conviction
on December 5, 2009.
(Franco Origlia/Getty Images)

met him at the concert hall. It made me miserable to know that being my boyfriend had cost him so much. On the other hand, I felt grateful that he, out of all the people in Perugia, was the person I was going through this with. Getting his first letter had renewed my faith in him, and we now wrote each other regularly. I knew I could trust Raffaele with my life. And I was.

Mignini and his co-prosecutor, Manuela Comodi, were determined to establish a connection between Guede, Raffaele, and me.

Their theory seemed to be that I knew Guede from the time Meredith and I had met with the guys downstairs in front of the fountain in Piazza IV Novembre—the night Guede told the guys I was cute. He hadn't made an impression on me at all then. The prosecution hypothesized that, after that night, he'd gotten in touch with me, perhaps about buying drugs. They stressed that we had a relationship, and although they allowed that it wasn't necessarily romantic, they insisted that Guede, like Raffaele, was obsessed with me. They further decided—based on a blurb Raffaele had written on his Facebook page way before he met me—that we'd been bored on the night of November 1 and headed to Piazza Grimana. There, Mignini said, we ran into Guede at the basketball court. I purportedly said, "Hey, let's go hang out at my place."

The prosecution spun this assumption further. According to Mignini, we found Meredith at the villa and said, Hey, that stupid bitch. Let's show Meredith. Let's get her to play a sex game.

I was horrified. *Who thinks like that?*

In their scenario, I hated Meredith because we'd argued about money. Hearing Mignini say that I told Guede to rape Meredith was upsetting. He added that I was the ringleader, telling Raffaele to hold her down. When he said that I threatened Meredith with a knife, I felt as if I'd been kicked. Even worse was hearing him say that when Meredith refused to have sex, I killed her.

When he initially said we were bored and went to the piazza, it sounded spontaneous, but now he said I'd tried to trap Meredith on Halloween—a holiday he saw as evil. Mignini based this on a text I'd sent Meredith asking her to hang out on Halloween. He emphasized Raffaele's and my supposed immorality and inclination toward violent fantasy. His "proof"? Raffaele's Japanese comic books about vampires and the one Marilyn Manson song he had downloaded. In closing arguments, Mignini said Meredith's murder was premeditated and was a rite celebrated on the occasion of the night of Halloween—a sexual and sacrificial ritual that, in the intention of the conspirators, should have occurred twenty-four hours earlier.

He was throwing motives against the wall to see which one stuck.

I wasn't used to the court lingo and depended on the young American woman who'd been appointed as my interpreter to fill me in on what was said.

The pretrial judge, Paolo Micheli, allowed testimony from two witnesses. The first was DNA analyst Patrizia Stefanoni for the Polizia Scientifica.

Starting right after we were indicted, Raffaele's and my lawyers had requested the raw data for all Stefanoni's forensic tests. How were the samples collected? How many cotton pads had her team used to swab the bathroom sink and the bidet? How often had they changed gloves? What tests had they done—and when? Which machines had they used, at what times, and on which days? What were the original unedited results of the DNA tests?

Her response was "No. We can't give you these documents you continue to ask for, because the ones you have will have to suffice."

Then, during pretrial, the defense lawyers pressed again, and this time Judge Micheli granted the request. Stefanoni gave us some documents—but not enough to interpret the data. When we objected, the judge shrugged and said, "Well, I asked her and she said those files aren't important for you."

Our only option was to question Stefanoni face-to-face about her methods.

Stefanoni was an attractive woman in her late thirties. Meticulously groomed, she had long dark hair, manicured nails, and olive skin. She wore tight suits that showed off her figure. The prosecution's questions were designed to let Stefanoni reassure the judge that all the testing had been done correctly.

Dr. Sarah Gino, our DNA expert, was a thin woman in her early thirties with short hair and thick glasses. Her convictions were absolute. She was insistent in her questioning of Stefanoni. Dr. Gino noted that Stefanoni hadn't provided enough information about her investigation for our defense to be able to critique her conclusions. How much of Meredith's DNA had she found on the knife blade? What was the evidence of the cleanup the prosecution was alleging?

Finally the judge granted Raffaele's and my defense teams' request for an independent review of Stefanoni's results. The expert he appointed was the head of the Polizia Scientifica—Stefanoni's boss.

Not surprisingly, this man rubber-stamped Stefanoni's work. It was done perfectly, he said.

That was the end of it.

The other testimony came from a witness named Hekuran

Kokomani, an Albanian man the prosecution called to prove that Raffaele and I both knew Rudy Guede. Our lawyers argued that Raffaele had never met Guede. I'd said "Hi" to him once when we hung out at the apartment downstairs. My other encounter with him was taking his drink order at Le Chic.

Kokomani said he'd seen the three of us together on Halloween, the day before the murder.

A massive lie.

Kokomani's testimony made the pretrial seem like a farce. According to him, after dinner on Halloween, driving along Viale Sant'Antonio, the busy thoroughfare just above our house, he came upon a black garbage bag in the middle of the road. When he got out of his car, he realized the "bag" was two people: Raffaele and me. He told the court that Raffaele punched him, and I pulled out a huge knife the length of a saber, lifting it high over my head. "Raffaele said, 'Don't worry about her. She's a girl,'" Kokomani testified. "Then I threw olives at her face."

As if this weren't nonsensical enough, next, Guede, whom Kokomani said he recognized from a bed-and-breakfast where he, Kokomani, worked, ambled up to the car. Kokomani said he asked Guede, "What's with the knife?"

"Guede said, 'Hey, brother, it's a party. We're just cutting some cake.'

"I know it was Amanda," Kokomani continued, "because I once met her uncle. Raffaele and Amanda were walking down the street. It was August. He offered me a beer." I wasn't in Perugia in August. I didn't know Raffaele yet. I don't have an uncle in Italy.

I was morbidly curious about Guede and simultaneously completely repulsed. Mostly I was disappointed. I had thought we'd have the chance to confront him. But he let his lawyers do all the talking.

Speaking for Guede, his lawyer, Walter Biscotti, claimed that his client was innocent. He said, essentially, "Rudy has told me his side of the story. What he says is not as outrageous as it seems. Rudy was at the villa because he and Meredith had agreed to meet and had fooled around before he went into the bathroom. He has indicated Knox and Sollecito are the actual perpetrators of the crime.

"Isn't that possible?" Biscotti asked. "Isn't that what the evidence shows? It shows him being there, and he's admitted to that. He says he left because he was scared. Of course he was scared! He's a young black man, living the best he could, abandoned by his parents. He stole sometimes, but out of necessity. I don't think there's enough evidence to say that he killed. The knife has Amanda's DNA, and the bra clasp has Raffaele's. Rudy admits that he was there, he tells what happened, and I believe him."

No witnesses were called for Guede. His lawyers could only interpret the evidence the prosecution had provided. They argued that his DNA had been found at the crime scene because he was scrambling to help Meredith and that he left because he was afraid. I remember his lawyer saying Guede didn't go to the disco to give himself an alibi but to let off steam. He escaped to Germany because he was worried that he'd be wrongly accused.

Biscotti was a little man who wore suits the color of . . . well, biscotti. His gray, curly hair sat like a thick rain cloud on his head. He came across as being convinced of the one thing no one else believed. He knew Rudy, Rudy was a good kid, and because he believed in his client's innocence, "Rudy should go free," he said.

The pretrial lasted for five and a half weeks—this wasn't surprising, because it was only held once or twice a week. Each time we met I would take the claustrophobic van ride, and when we arrived I would have the stomach-twisting encounter with the press. Even if I understood Italian pretty well by then, the technical discussions about DNA would have been impenetrable for me. I was weary of this soul-killing routine. It seemed to me that it was time for me to be freed and go home.

My family was as optimistic as I was. I had been in prison for about a year, and the anniversary seemed an appropriate time for my release. But the closer we got to the end, the more pessimistic my lawyers grew. "The judge will probably rule that you'll have to be tried for murder, because there's so much attention on this case," Carlo said. "We have to be ready for that."

This broke my heart. "Why can't everything be resolved in the pretrial?"

"It's more complicated than that, Amanda. It's likely going to take a jury. We will have to call witnesses."

But Luciano and Carlo never completely lost hope. "It's possible this could end well," Carlo said. I clung to that chance.

Still, there were reasons to be worried. Because the prosecution was withholding information, there was evidence I couldn't refute: the knife, my "bloody" footprints, Raffaele's DNA on Meredith's bra clasp. And how would we fight the prosecution's claim that we'd cleaned up the crime scene? I went to sleep every night telling myself that it would work out because we were innocent—and because it was so clear that Guede was guilty and lying.

My lawyers argued exhaustively that Meredith and I had been friends—that there was no animosity between us. They argued that we had no connection to Guede, that Kokomani was a lunatic. But the case hinged on DNA, not on logic.

On October 28, the final day, I got to speak for myself. Since the judge understood English, I stood up without my interpreter and tried to explain what had happened during my interrogation. I told the judge that I hadn't meant to name Patrick or to cause confusion but that the interrogation had been the most brutish, terrifying experience of my life. I'd been exhausted to begin with, and I had gotten so scared and confused that it was as though I went out of my mind. My interrogators told me that they had evidence I'd been at the villa, that Raffaele was no longer vouching for my whereabouts that night, that I had been through such a horrible trauma, I had amnesia. "I believed them! I'm innocent!" I cried.

I was shaking, so nervous I couldn't go on. I didn't mean to cry, but I wept uncontrollably. Recalling the interrogation struck me at my core. I was so eager for a positive decision, so eager for this ordeal to be over, that I couldn't keep myself together. Afterward, I hunched over in my chair, ashamed of having lost hold of my emotions. Luciano and Carlo patted my head and rubbed my back. "Don't worry," Carlo insisted. "You did fine."

When the prosecution rested their case, Mignini demanded a life sentence for Guede and a full trial for Raffaele and me.

After the judge retired to his chambers, we were each taken

to a different empty office in the courthouse to wait for his decision. Raffaele folded a page from that day's newspaper into a flower, which the guards brought to me. But I was focused on Guede, who was being held in the room next to mine. I could hear him talking with the guards, cracking jokes, and chuckling. I was fuming! I wanted to beat on the wall and tell him to shut up. His nonchalance incensed me. I thought, *Does no one else feel this?*

Six hours into what would be a day-long wait, I got to see my lawyers. "It's unusual for the judge to deliberate this long," Carlo said excitedly. "If the decision were easy, it would have happened already."

Carlo's optimism fueled my own. Maybe the judge will be gutsy enough to see how preposterous Hekuran Kokomani's wild story was and the truth in Sarah Gino's questions. *Oh God, this is taking forever. It must be a good sign.*

Late Tuesday afternoon, when the sky was dark, word came up. The judge was ready.

I entered the courtroom. I could barely walk.

Judge Micheli read Guede's verdict first: Guilty for the sexual assault and murder of Meredith Kercher, with a sentence of thirty years.

The verdict didn't surprise me at all—for a second, I was enormously relieved. I thought, *He's the one who did it.* The judge's delivery was so flat he could have been reading the ingredients off a box of bran flakes. Still, my chest clenched when I heard "thirty years." Not because I pitied Guede. I'd been so focused on whether he would be found guilty or innocent, I hadn't thought about the length of his sentence. I was twenty-one; thirty years was more time than I'd been alive—by a lot.

I breathed in.

"The court orders that Knox, Amanda, and Sollecito, Raffaele, be sent to trial."

I broke down in huge, gulping sobs. I'd made a heartfelt plea— "I'm telling you I'm innocent! I'm sorry for any of the confusion I've contributed."

The judge hadn't believed me.

On the heels of the announcement "This court is adjourned," the guards started walking me out.

"Amanda, don't cry," Carlo called. "Don't worry. This isn't the end. The trial is different from pretrial. Both sides get to put forth witnesses. We're going to analyze everything. We're going to prove them wrong."

I wanted him to be right. But all I could think was if the court hadn't believed me this time, why would they believe me the next time?

Chapter 24

———————

October–December 2008

I *n prison there is never just one stress.*
      That phrase was becoming my mantra.
      About a week before my pretrial ended, another trau-
matic episode began. One gray day, when I was outside exercis-
ing, I saw some prisoners I had become friends with. "Hey!" I
yelled. "Good morning!"

They glared at me but said nothing. I walked away.

Something is wrong, but what? What's going on?

After walking a couple of laps by myself, even my headphones
didn't block out the painful silence. Tears started rolling down
my cheeks. I was too upset to keep going, but I was afraid to stop
for fear that whatever was happening would catch me as soon as
I stood still.

Finally the guards called me to the office of the *ispettore*, a
round middle-aged woman with short, wispy orange-dyed hair. I
found her standing behind her desk pointing at a copy of *Corriere
dell'Umbria*, the local paper, spread open in front of her. "How do
you explain this?" she demanded.

"*Spiegare che cosa?*" I asked, baffled. "Explain what?"

I could see that the headline said something about me.

"It's an interview," she said. "It talks about Cera."

"You know I don't give interviews!" I said.

The inspector turned the paper around so I could read the article. The reporter claimed to have interviewed my mother, who talked about things I'd said.

"You need to tell your mother to refrain from speaking about the inner workings of the prison," the *ispettore* said sternly.

"My mom would never do that!" I screeched. "She only gives interviews to talk about my innocence. She would never reveal our private conversations."

But the article was full of insider information. They'd gotten Cera's name and certain details right. They said she kissed me once and that I feared further sexual harassment. They knew she was a cleaning fanatic and that she wouldn't let me make coffee because it would leave water spots on the sink.

Now I knew why my prison friends were shunning me. Like Wilma, I was now an *infame*. The *ispettore*, guards, and prisoners assumed I'd told my mom to tell journalists that I was being harassed and abused in prison, betraying Cera to gain public sympathy.

"Maybe a guard talked," I said to the *ispettore*.

She scowled at me.

Who, other than a few guards and my family, has access to my conversations?

My lawyers later explained. "Remember, the conversations you had with your parents were bugged," Carlo said. The prosecution entered the transcriptions of the conversations as evidence, which was why they were made public.

By the time I knew this, though, no prisoner was talking to

me. I willed myself not to care. They wouldn't have listened to an explanation anyway. Now, both inside and outside, I was being accused of something I hadn't done.

Cera had been the one to tell me how mean, how crazy, how awful, prisoners could be to one another. I hadn't wanted to believe her, and I'd promised myself that I'd never become bitter like she was. But I was getting closer. I refused to become so cynical and angry that I felt spite, but my natural hopefulness was flagging.

Even though I was no longer separated from the rest of the prisoners, as I had been for months, I felt more isolated than ever. The few prisoners who did acknowledge me glowered.

Only Fanta, the young Roma woman who delivered groceries, said hello, and she'd often stop by my cell and tell me jokes.

Prison is a hard, raw place, where people think of themselves before others and where compassion is often forsaken.

Don Saulo was the one person who cared about any of us. In spite of the awful way the other prisoners treated me, he restored some of my faith in humankind. "It doesn't matter what people think you did," he told me. "What matters is what you did do. Don't worry if people can't see your goodness. The only important thing is your conscience. You have to take heart and strength in that."

Happily for me, my stepfather Chris's job let him telecommute from Perugia. His advice about standing up to the other prisoners was good, if not practical. And it made me laugh. "You need to grow some big *cojones*," he said. "Yours are a little too small. You need some real big fat ones," he said, making a squeezing gesture with his hands.

We held onto the belief that the law would be on my side when my trial started. I was innocent. No matter how the prosecution misconstrued things, there would never be evidence enough to convict me. And I had the great consolation of knowing that prison wasn't my world. In time, I'd be set free. I could survive this as long as it took. But I never thought it would take years.

The other person who gave me hope during this time was an Italian professor from UW. I hadn't studied under him yet, but he organized an independent study class for me in which I got to read, write, and translate Italian poetry and short stories. I was frustrated by the academic time I'd lost, and I was determined not to waste another minute. *You came to Italy to learn Italian, Amanda,* I told myself. *Immerse yourself in it 24/7.*

I kept saying good morning to the other prisoners. In time, some returned a curt *"Ciao."* Most didn't. Cera had told them to ignore me, and for their own preservation, they did.

The only place I found peace was inside my own head. I started expecting nothing. The one thing that surprised me was the occasional time another prisoner, like Fanta, treated me kindly. As excruciating as this was, it forced me to develop a sense of independence, a faith in myself.

Chapter 25

January–March 2009

T he pretrial had been like the first reading of a play. No costumes, no audience, no reporters, and very few players. It was held in chambers and closed to the press. The lawyers wore suits. Only two witnesses—the prosecution's DNA analyst and a man who claimed to have seen Rudy Guede, Raffaele, and me together—testified.

The full trial for Raffaele and me was like opening night. I wasn't prepared for the spectacle. We were tried in Perugia's fifteenth-century courthouse, in a large courtroom known as the Hall of Frescoes—L'Aula degli Affreschi. The walls were stone, the windows spanned from floor to vaulted ceiling. A giant crucifix hung behind the bench. The two judges, two prosecutors, and eight lawyers wore black robes with lacy collars. The six computer-selected jurors, all middle-aged, wore sashes in the green, white, and red of the Italian flag. The trial was open to the press, who were more than a hundred strong.

Three no-nonsense guards—one in front of me and one on either side—led me in through the door in the back of the packed courtroom. Police officers, including some who had interrogated me fourteen months before, were lined up against the back wall. I knew that almost every observer thought I was guilty and wanted me to suffer.

A fenced-off area separated the spectators, journalists, TV cameras, and photographers from the defense and prosecution teams, including the two of us on trial. The press snapped pictures and yelled in English and Italian, "Amanda, Amanda, what do you have to say?"

I exhaled as I walked past Raffaele's family and my own. Sitting behind the defense tables and in front of the media, they were the only friendly faces in the room. Mom wasn't allowed inside until after she testified, but seeing my aunt and uncle, Christina and Kevin, who had made the trip to Perugia in place of Dad and Chris, filled me with gratitude. I knew I wasn't alone. I gave them a little wave and a big smile to let them know how glad I was they were there. I never anticipated that that smile would be reported as "Amanda Knox beamed as she was led into an Italian court." And the *Daily Mail* amped up my regular walk: "She made her entrance like a Hollywood diva sashaying along the red carpet." I don't know if the reporting was skewed to sell papers or if the presumption of my guilt colored the way the reporters saw me. Anyone reading or watching the TV reports would have come away believing the girl called Foxy Knoxy was amoral, psychotic, and depraved.

At one end of the room was a black metal cage used to hold dangerous criminals. I thought, *Oh God, they're going to put me in that cage*. It wasn't rational—but my anxiety was ratcheted up to maximum. I was terrified. I felt paralyzed.

To my tremendous relief, the guards steered me to the table where Carlo and Luciano were sitting. My lawyers and I were on the far right, with Raffaele less than ten feet away, at the next table. The prosecution sat on the far left, with the civil attorneys for the Kerchers and Patrick at a table behind them.

In the United States, civil and criminal trials are held separately; in Italy, they're combined. The Italians clearly believe their jurors can compartmentalize—the same eight people decide all the verdicts. Moreover, jury members are not screened for bias, nor guarded from outside influence. The government was trying Raffaele and me for five crimes: murder, illegally carrying a knife, rape, theft, simulating a robbery, and a sixth just for me: slander. The Kerchers, believing Raffaele and I had killed their daughter, were suing both of us for €5 million—about $6.4 million—€1 million for each of Meredith's five family members, to compensate for their loss and emotional anguish. Patrick Lumumba was suing me for slander for a yet to be determined amount. The owner of the villa was suing me for €10,000 for damages and lost rent.

Some evidence, including my 5:45 A.M. "confession," when I confusedly described Patrick as the murderer, wasn't allowed to be introduced in the criminal case. At that moment I had already officially became a suspect and had a right to a lawyer. The same evidence could be, and was, discussed in front of the jury in the civil cases.

The way the Italian justice system works is that during deliberations, each of the judges and jurors gets to say what he or she believes the sentence should be—from nothing to life imprisonment. Unlike in the United States, where the decision has to be unanimous, what's required in Italy is a majority consensus—the maximum sentence supported by at least five jurors.

I sank into a big, plush chair sandwiched between Carlo and

my court-appointed interpreter, who'd been brought in because I was being tried in Italian. We stood as soon as the court secretary summoned the court to order, announcing, *"La corte"*— "the court." I was thankful that the arrival of the judges and the jury took the focus off me.

I began the trial with mixed emotions. The pretrial had so squashed my hope of being released that I dreaded what could come next. But my intense natural optimism, unhampered by logic or media predictions, helped ease my despair.

Carlo and Luciano had prepped me for a long trial, but I discounted that, too. Surely the trial would be speedy, and Raffaele and I would be found innocent, because we *were* innocent. And the court was obligated to be just. I'd spent fourteen months in prison. I couldn't allow myself to believe that I'd spend several more bouncing between courtroom and cell.

I was naïve.

It took hearing only a few sentences for me to know that the interpreter was giving me the condensed version. The one plus to prison was that my Italian had improved so much that I could think in the language. I decided not to use her anymore. My lawyers could explain what I didn't understand.

The first thing discussed was the motive. The prosecution's simple story was absolutely false, but it apparently rang true for the authorities. They added flourishes in the course of the trial— Meredith was smarter, prettier, more popular, neater, and less into drugs and sex than I was. For some of or all these reasons, she was a better person, and I, unable to compete, had hated her for it. I had cut her throat in rage and revenge. It was idiotic.

Mignini relied heavily on the testimony of Meredith's British girlfriends. Robyn Butterworth testified that my unconventional behavior had made Meredith uneasy. The others agreed—they said I brought male friends over, didn't know to use the toilet brush, and was too out in the open about sex. Small details built up to become towering walls that my defense team couldn't scale. I was done in by a prank gift and my unfamiliarity with Italian plumbing.

Questioning Robyn, Mignini said, "Do you remember if Meredith said Amanda left certain objects in the bathroom?"

"Yes, actually I saw these objects myself," Robyn responded. "In the bathroom there was a beauty case with condoms and a vibrator and other objects. Meredith told us it was a little strange, she felt uncomfortable, because Amanda had left them where anyone could see them."

"So it was Meredith who told you what was in the beauty case?" Mignini prompted her.

"Yes," Robyn answered.

"Was Meredith irritated with Amanda over this?"

"Not really, but it seemed a little strange; it seemed strange to her more than anything else."

I was making notes when each witness spoke, to help prepare Carlo or Luciano for his cross-examination. I'd scribble comments like "That's not true" or "I don't get what she's saying, because it didn't happen the way she's describing." Then my lawyers would weave that into their questioning. They'd explained before the trial began that I was allowed to make spontaneous declarations but had asked me to trust them to get our points across instead of interrupting the court proceedings. They reminded me that I'd have my day when I testified.

My frustration doubled when Robyn talked about the bunny vibrator. I had to clarify this. When Brett gave it to me, TV shows like *Friends* and *Sex and the City* were an American obsession, with characters using vibrators as gags. The prosecution put the emphasis on sex—and me. The vibrator was proof that I was sex-obsessed—and proof that my behavior had bothered Meredith.

I leaned over to Luciano. "I want to say something," I whispered.

Luciano cleared his throat. "My client would like to make a spontaneous declaration," he announced.

"Please go ahead," the judge directed me.

I stood. "Good morning, Judge," I began. I was suddenly burning up, even on that cold February day. "I want to briefly clarify this question of the beauty case that should still be in my bathroom. This vibrator exists. It was a joke, a gift from a girlfriend before I arrived in Italy. It's a little pink bunny about this long . . ."

I held up my thumb and index finger to demonstrate.

"About this long?" Judge Giancarlo Massei said, holding up two fingers to clarify.

"Yes," I said, turning red with embarrassment.

"Ten centimeters [four inches]," he said for the court record.

"I also want to say that I'm innocent, and I trust that everything will come out, that everything will work out. Thank you."

I remember thinking while I was speaking, *Oh my God, I hope I don't sound as stupid as I think I do.* I sat down fast.

I wasn't making excuses for the vibrator. I just wanted to put it into perspective—that it was a gag gift, not to be taken seriously, and that Meredith had never complained about it to me.

In my rush to explain and my uncertainty over what I was allowed to bring up, I didn't stop to think that this was the first time most of the people in the courtroom would be hearing me speak. I should have clarified which friends I'd invited to the house—I had sex at the villa with only one guy. The rest of the friends I'd brought over were just that, friends, and I didn't bring them over in the middle of the night. Most important, I should have talked about my friendship with Meredith.

I didn't know yet that I was allowed to contradict witnesses whose testimony was wrong.

Now, instead of dispelling the notion of me as a sex fiend, I had burned it into the jury's and the public's consciousness.

In my flustered state, the only thing I did well was express my faith in the court. My lawyers had told me that I couldn't hope for justice for myself if I appeared to distrust the Italian legal system.

I wasn't sure I had faith in Italian justice, but what choice did I have? I had to believe things were going to turn out well for me.

It did seem I'd won a small victory when Mignini questioned my former housemate Filomena. She insisted that Meredith and I got along fine and hadn't had a falling-out—only that we'd "developed different personal interests." She didn't make a big deal over the friends I brought home.

Other parts of Filomena's testimony irked me. When Mignini asked how we divided up chores in the villa, she said that we took turns. "Turns were not always respected," she added.

"Who didn't respect them?" Mignini asked.

"Amanda a few times didn't respect them," Filomena replied.

Filomena can't be saying this, I thought, straining not to blurt out my disbelief. Laura had drawn up a cleaning chart only a couple of days before Meredith's murder—my day hadn't even come up yet—and the prosecutor was trying to build a case that I was careless and inconsiderate. Filomena must have thought I was slack about cleaning, and this never-before-stated resentment hurt my feelings.

Because of our age difference and the language barrier, Filomena was the housemate I knew the least. In the short time we'd lived together, she'd acted big-sisterly toward me, and we'd gotten along well. I'd felt truly content when my three housemates and I had sat around with the guys from downstairs after lunch or dinner, passing a joint, chatting, and laughing. Smoking pot was one of the ways we socialized together. But when Raffaele's lawyer Luca Maori cross-examined her about her drug use, Filomena rewrote our shared history. "To tell you the truth, I sinned once," she said, looking down at her lap. "I sinned."

I felt a stab of anger.

"We are all sinners," Maori sympathized.

"I sinned," Filomena repeated.

"So you've used it once?" Maori asked.

"Yes."

It was painful for me to realize that Filomena seemed to care more about her reputation than about how her insincerity would reflect on me as I stood trial for murder. Laura and Filomena had always bought the marijuana for the villa's personal use. But when Filomena shrugged her shoulders helplessly on the stand, she made it seem that the only reason marijuana was in the house was because of me.

What bothered me most wasn't what she said. I watched her carefully the whole time she was testifying. Whether it was because she thought I was guilty or because she felt ashamed for what she'd said, she never looked at me.

During her testimony a week later, Laura also avoided eye

contact—and it was every bit as hurtful. But I was pleased that, at least under questioning, she didn't make it seem that my behavior had been out of step with the rest of the house. When Mignini brought up names of guys who'd come over, Laura replied, "Those are my friends." When he asked if anyone in the villa smoked marijuana, she said, "Everyone."

Then the prosecutor mentioned the hickey Raffaele had given me when we were fooling around the night of November 1. "Did you see if Amanda had an injury, a scratch, some wound?" he asked her.

"I noticed that Amanda had a wound on her neck when we were in the *questura*," Laura answered, "precisely because Meredith had been killed with a cut to her neck. I was afraid that Amanda, too, might have been wounded."

I liked Laura and had looked up to her. She'd lent me her guitar and thought it was cool that I practiced yoga. There was only one reason why she would turn a love bite into a sign of my involvement in the murder. My stomach plunged to my knees. *I can't believe Laura, of all people, thinks I'm guilty.*

I felt completely betrayed.

Even though my speaking up over the vibrator had been a disaster, I couldn't keep myself from addressing Laura and Filomena's testimony. So I asked to make another spontaneous declaration.

"Thank you, Your Honor," I said. "It truly and sincerely troubled me to hear that after all this time there's a certain exaggeration about the cleaning. This was an absolute exaggeration. It wasn't a thing of conflict. Never. In fact I always had a good relationship with these people. This is why I'm truly troubled—troubled, because it wasn't like that. So, thank you."

Hours of watching TV courtroom dramas hadn't prepared me for how tense and uncomfortable an actual courtroom can be—or how defendants in the real world don't have a script handed to them. Standing there with the eyes of the world on me—my every word, gesture, and inflection scrutinized—I might as well have been naked in Piazza IV Novembre.

Still, I wished I'd pushed my lawyers to let me speak more often. Luciano and Carlo's intentions were good, but I believe they underestimated the power of my voice and the damaging effect of my silence.

Even with my clumsy efforts to defend myself—and with other people describing me as the girl with a vibrator, a slob, a girl with a "scratch" on her neck—what did the most damage in those early weeks was a simple T-shirt, and that was my own fault.

My stepmother, Cassandra, had sent me the shirt. It had fat, six-inch-tall pink lettering that blared "All You Need Is Love"—a line from one of my favorite Beatles songs. I loved it, and I wore it on Valentine's Day—the day Laura testified.

When I passed the press pit on the way in and out each day, I never paid the journalists or photographers any attention. When my parents visited me, they'd fill me in on what "those idiot journalists" were writing about. "They aren't talking about the cross-examination. They're only talking about your hair," my dad said more than once.

Luciano and Carlo had said that what mattered was what happened in court—that we had to show that the prosecution was wrong. "You're a good girl. Just be who you are," Luciano said.

One person, trying to be helpful, suggested that I wear a cross on a chain around my neck to court. I rejected that outright. I couldn't pretend I'd found religion.

I thought that if I dressed in my usual jeans and a T-shirt, the judges and jury would see me for who I really was, not as Foxy Knoxy, not as someone who was dressing to impress the authorities.

I'm glad I didn't wear a cross, but in hindsight I do wish someone had told me that my clothes should reflect the seriousness of the setting and my situation—that they were another way to convey my respect to the court.

So when I wore the "All You Need Is Love" T-shirt, the press dwelled on what I *meant* by it. Is Amanda trying to say all she needs is love from the jury? One British newspaper headlined its story about that day's hearing, "Obnoxious: Murder Trial Girl's Love-Slogan T-Shirt. "Knox's narcissistic pleasure at catching the eye of the media and her apparent nonchalant attitude during most of the proceedings show the signs of a psychopathic personality," the article said.

I felt foolish after the T-shirt episode. I never again wore anything that might be seen as attention-grabbing. The press still commented on my clothes, my hair, and whether I was happy, sad, tearful, bored. Their zoom lenses tried to capture what I wrote on my notepad.

The press wrote that I had to be the center of attention. In reality, prison had taught me I was nothing. Nothing revolved around me. Nothing I said mattered. I had no power. I was just occupying space. I wanted to disappear. I didn't want to be me anymore.

———————

The hearing days were often exhausting—as were our off days.

For the first few months of the trial, I was still an *infame*, with no one among the prisoners but Fanta, whom I'd since moved in with, to talk to. Then, early one morning, without warning, Cera was told that she had a half hour to gather her things. She was being transferred to a prison in Rome. By the time the prison day started, she had already left. I felt sorry for her—she had made a sort of home for herself at Capanne.

I was also sorry that we had never managed to make up, to be on good terms, even talking terms, with each other again. She had taught me a lot about prison, and had ultimately seemed more damaged than bad.

Later that same morning, at nine, when I went outside for the first *passeggio*, everyone I passed as I walked my laps said, "Good morning."

Just like that! The silent treatment ended as abruptly as it had begun.

One prisoner who had been at Capanne for months but whom I'd never met came up to me and introduced herself. "I'm Dura," she said. "I always knew you weren't as bad as Cera said you were. But what could I do about it?"

It helped me that the tension I'd been living under dissipated somewhat, but not much besides that changed for me. I had found I liked my own company, and I went on keeping to myself.

What didn't get easier was room sharing. In part this was because of turnover. The prison population had doubled in the months since I'd been at Capanne. They couldn't add rooms, so they added beds. Mona arrived soon after a fifth bed had been jammed atop one of the beds as a bunk in our four-person cell. She was a heavyset, butch woman from Naples with scars from cuts on her arms and several missing teeth. Mona was in her thirties and totally unpredictable—as happy and well behaved as a child on Christmas Eve or as blindly furious as a rampaging bull. Always heavily medicated, she slept a lot and was disoriented when awake.

The four of us were unhappy about taking on a fifth roommate, and the fifth was furious that she'd been transferred. Before settling in, she hurled a stool across the room, screaming for the *agente*.

During *socialità* she hooked up with a more feminine-looking prisoner named Gaetana, but their affair was short-lived. Gaetana dumped Mona for another masculine woman from Southern Italy, and the new couple often sat together on a bench outside, hands pressed on each other's knees.

One day, during *passeggio*, I walked in front of Gaetana's bench. Mona charged at the couple. In the way, I jumped to the side seconds before Mona fell on Gaetana and her girlfriend, punching both in the face. The attack escalated into a full-scale brawl before the other prisoners managed to pull Mona off. I sat cringing in a corner of the courtyard, too shaken to stand up. I still hadn't gotten used to the fact that violence could erupt anywhere, anytime.

Once, after I'd straightened up our cell, Mona stormed up to me. "Amanda, where's my tobacco?"

Her anger made me skittish. "I didn't see it while I was cleaning," I said.

"I had it by my bed, and now it's gone. Don't play games with me," she hissed, balling her fists.

"I swear I didn't see it," I said.

"I found it!" Fanta called from the bathroom. She walked into the main room carrying a wadded scrap of paper. Mona grabbed it, opening it to reveal loose tobacco. "Amanda must have accidentally thrown it away thinking it was trash," Fanta said.

Mona spun around, facing me. "Are you trying to make a fool out of me?" she thundered. "You lied to me!"

"I didn't realize! I'm sorry!" I cried, inching backward.

"Mona! Mona! It's okay. It was a mistake!" Fanta said, stepping between us.

I don't know what would have happened if Fanta hadn't been there.

The next cellmate was Ossa; she moved in when Mona moved out. A wiry, young Roma a year older than me, she was strictly religious and as sulky as a teenager.

We started out on good terms. When Ossa first arrived, my other cellmates didn't like her because she slept almost all day. I argued that she had the right to sleep if she wanted—she wasn't in anyone's way. She talked to me about God, speaking in tongues, how she was like Robin Hood—stealing from the rich to give to the poor.

One day Ossa told us she'd asked an inmate she'd become friends with—a woman who'd been caught stealing from other prisoners—to take a newly empty bed in our cell. Two of the four of us would have to sign off on the official request form for the *ispettore* to grant the move.

We all said no.

Ossa took out her resentment on me. She scoffed at me for reading and writing, and for turning the TV volume down. She was disdainful of the groceries I ordered—and bought—for everyone in the cell. She sarcastically called me "queen of the cell." Whenever my case came on the news, she'd agree with the prosecution. "Everyone knows you're guilty, Amanda—and fake," she said once. Fuming, I buried my face in my book, *The Ultimate Hitchhiker's Guide to the Galaxy*, not wanting her to know she was getting to me.

My indifference so infuriated her she threatened to flush my head in the toilet.

One night, when I was sitting on the floor by the cell door trying to catch the light from the hallway in order to read, she leapt out of bed and lunged at me, screaming, "I hate you!" Her hands were raised to slap me when another cellmate, Tanya, jumped up and grabbed Ossa from behind, pinning her arms back. Immobilized by fear, I felt like a roach about to be squashed. While on trial, I couldn't afford a mark on my record, especially not for fighting. Shaking, I stood with my arms up in surrender. "Even if you hit me," I said softly, "I'm not going to hit you back."

The next day, Ossa watched while I wrote in my diary and then stepped up to my bed. "You're writing horrible things about me in there!" she shouted, grabbing the notebook off my lap. Before I could react, she'd ripped out a huge chunk, tearing the pages to shreds.

I felt completely helpless. *If I try to grab it back, she'll tear out my hair.* This felt almost as invasive as when the police confiscated my journal soon after my arrest. The effect was the same: I had no choice but to stand by, paralyzed, as I lost something that was worthless to anyone but me, and was the possession I most cared about: my thoughts.

A few days later, to my great relief, Ossa was freed. Gone from my life just like Cera.

———

The weeks passed, and as the prosecution called a long line of witnesses, my optimism over a speedy acquittal was shrinking. Helpless sadness took over.

I expected the prosecution to call police officers who'd been at the villa and those who were in the interrogation room, but initially I didn't recognize Officer Monica Napoleoni. I'd never seen her dressed to suit her title—head of the Division for Homicide Investigation. Usually she wore skin-tight jeans, form-fitting shirts, and flashy sunglasses. Wearing a dark blue jacket adorned with medals the size of silver dollars, she now looked so unlike herself that it seemed she was playing dress-up to convince people of her authority. Everything she did and said—her choice of words, the content, and the emphasis—was to impress the judges and jury with her professionalism. She defended the shoddy work of her investigators. She was repellent. She was in control of herself, sitting in a court of law and lying without a second's hesitation.

When she answered Prosecutor Mignini's questions, she was clear, straightforward, and self-serving. She was smarter than her fellow officers. She knew the court was looking for police slip-ups. "We did our jobs perfectly, all the time," she testified. "We didn't hit Amanda." "We're the good guys."

When the defense questioned her, Napoleoni's manner switched from professional—albeit dishonest—to exasperated, incredulous, and condescending. For instance, when Raffaele's lawyer Giulia Bongiorno asked if the gloves police used at the crime scene were sterilized or one-use gloves, Napoleoni took a snarky tone, saying, "It's the same thing."

"By one-use gloves you mean that they are gloves that can be used only once, right?" Bongiorno asked.

"Obviously, yes," Napoleoni said haughtily.

"Therefore it means that every time you touched an object you changed gloves?"

"No, it means that I put them on when I enter before I touch objects, and that's what I did."

"But therefore with the same gloves, without changing gloves, you touched the various objects in the room in the course of the search?" Bongiorno asked.

"It's obvious, yes."

I knew it was the police's job to analyze the scene of a crime, gather clues, and determine who did it. But here in Perugia the police and the prosecutor seemed to be coming at Meredith's murder from the opposite direction. The investigation was *sospettocentrico*—"suspect-oriented": they decided almost instantly that Raffaele and I were guilty and then made the clues fit their theory. Instead of impartiality, the prosecution's forensic experts were relentless in their drive to incriminate us. Their campaign was astonishing for its brashness and its singleness of purpose.

Napoleoni built up her case against Raffaele and me by tearing us down. The police were suspicious of us from the start, she said, referring to the first time she saw us, in front of the villa on the day Meredith's body was discovered. "Amanda Knox and Raffaele Sollecito kept their distance from the others, kissing, caressing each other," she said.

Napoleoni added that, later, at the *questura*, we "were absolutely indifferent to everyone. They sprawled in the waiting room, sprawled on the seats, kissed each other, made faces at each other the entire time . . . They talked to each other under their breath. I noted their behavior because it seemed impossible that these two kids thought to kiss each other when the body of their friend had been found in those conditions."

My housemates and their friends reacted more appropriately, Napoleoni said. They "were all crying," she told the court. "Some despaired."

To Napoleoni, Raffaele and I were self-centered narcissists.

We lacked basic compassion. And we were liars through and through.

Even something as simple as a misunderstanding about the toilet in the villa fueled the police's suspicions. Napoleoni explained that I'd said that morning that when I came to the villa to shower, I'd "noticed feces in the toilet." But when I returned with Raffaele, I "didn't find it there."

That was true.

However, when Napoleoni checked, she said, "I saw that the waste was very evident. The smear started from the top of the bowl. I went out again and told them, 'but no, it's there,' and they began to fall into contradiction. Something wasn't right."

I was surprised but didn't doubt her. Realizing that someone had broken in, I'd been afraid when I went back in the villa with Raffaele. I looked at the toilet from a distance and, not seeing anything in the bowl, assumed someone had flushed it. Clearly, I was wrong. Apparently the feces had slid down farther into the bowl. But Napoleoni acted as if, in discovering the unflushed toilet, she'd caught us in a lie and that we'd ineptly scrambled to come up with a cover.

At the time, it seemed urgent to tell Napoleoni about the unflushed toilet. I thought it was important for the police to know that the killer might have been in the house when I came home the first time. Why would I make up a story about disappearing shit?

Napoleoni went on, twisting each aspect of the case. "I immediately noted that the house couldn't have been broken into from the outside. It seemed to have been done after the room was made a mess. I immediately noted that there was glass on the windowsill, and if a stone came from the outside, the glass should have fallen below."

She also said that when the Postal Police came to the villa with the phones Meredith had been using, "they asked Amanda if it was normal that Meredith locked her door. Amanda said Meredith always locked her door, even when taking a shower."

Filomena Romanelli disputed this, Napoleoni said.

What I'd said—that Meredith sometimes locked her door, including sometimes after she showered, while she was changing, and when she went out of town for the weekend—had gotten garbled in translation. The mistake cost me credibility. Having caught me in what they took to be small lies, the cops saw me as someone incapable of telling the truth.

The homicide chief added that by checking telephone activity tables, the police discovered that both my cell phone and Raffaele's had been inactive the night before Meredith was found. "Amanda from 8:35 P.M. and Sollecito from 8:42 P.M." That fact meant nothing, but Napoleoni presented it as if, in turning off our phones, we had had an ulterior motive. That we'd wanted to watch a movie without being interrupted did not come up.

"We looked for contradictions," Napoleoni told the court, "and the contradictions always came from Amanda and Raffaele, because the account they gave us was too strange. It was improbable."

If anything, it was surreal. I hadn't expected to come home to a murder scene. I hadn't known what to make of what I'd found. Yes, I'd come home and taken a shower. I didn't investigate beyond Meredith's closed door. And then one thing had led to another. I'd discovered droplets of blood, then an unflushed toilet, then a break-in in Filomena's room, and finally the police found Meredith's body.

Because Raffaele and I reacted differently from the others—

and, I assume, differently from how Napoleoni imagined she would have—she and the prosecutor decided that Raffaele and I were the killers. Of course it's natural for people to jump to conclusions, but not for a police officer to ignore facts and rely on superficial impressions. My stomach burned with resentment. I wanted to shout at her, "Who says there's only one way to react? Who decided that being different equals being guilty?"

When I first met Napoleoni, I thought she was mean. When I spent more time around her, I thought she was hateful. But looking at her on the stand, I thought, *You were so stupid, Amanda. How could you not have realized that Napoleoni pegged you for guilty from the start?*

I remembered sitting in the back of the squad car on November 3, when the police were driving me to the house. Napoleoni was in the passenger seat in front. I said I was tired. She swiveled around to glare at me. "Do you think we're not tired? We're working 24/7 to solve this crime, and you need to stop complaining," she reprimanded me sharply. "Do you just not care that someone murdered your friend?"

I felt put upon that day. The police were guilt-tripping me. They didn't understand that my life had been shattered. They were used to the stress of their work, but I think they didn't realize that regular people get tired, hungry, and overwhelmed.

I was also frustrated with myself. I couldn't seem to do anything right.

In the Hall of Frescoes, the authorities made the same points as Napoleoni, one after another, often using the same words: I was strange, my behavior suspicious.

You could tell their testimony had been rehearsed.

On the stand, my chief interrogator, Rita Ficarra, seemed

much smaller than she had at the police station. Middle-aged, with dull, shoulder-length brown hair, she came across as reasonable. Who would believe that she'd been ruthless, questioning me for hours, refusing to believe that I didn't know who'd murdered Meredith? I wondered how this woman, who now struck me as average in every way, had instilled such fear in me.

Like Napoleoni, Ficarra insisted, "No one hit her." She was serene and straight-faced as she testified.

Ficarra elaborated. "Everyone treated her nicely. We gave her tea. I myself brought her down to get something to eat in the morning," she said, as if she were the host at a B&B. Then she added, "She was the one who came in and started acting weird, accusing people."

In her story, I was the crazy guest.

When Raffaele was called to the *questura* on November 5, I went along because I was afraid to stay at his apartment alone. Ficarra's take on this was not generous: "She just came in," she said. "No one called her in."

She told the jury that when she had returned to the *questura* at around 11 P.M., she and her colleague came through the door and into the hall. "I found Amanda . . . My astonishment was that I found her demonstrating her gymnastic abilities. She did a cartwheel, a bridge, she did splits," Ficarra said. "It honestly seemed out of place to me."

Ficarra didn't mention that the silver-haired police officer had asked me to show him how flexible I was. Now I can't believe I acquiesced to his request, that it was normal to do yoga in such a setting.

The longer Ficarra testified, the more she made it seem that the pressure the police exerted on me to confess was all in my head, that I'd blown the interrogation out of proportion. "In the end it was a calm dialogue, because I tried to make her understand that our intent was to seek collaboration," she said.

"At first she denied being at the villa the night of the murder, and then, when we called her on it, she started blaming someone else."

It was nearly unbearable to listen to her describe their behavior toward me as gentle and considerate. She defended everything without flinching. It was all I could do not to jump up and scream, "No! That's not at all how it happened!" But my lawyers strongly advised me not to say anything—that as I was someone who had already been accused of lying, no one would take my word, especially over the police's.

Judge Massei asked Ficarra if I spoke to her in English or Italian.

"In Italian," Ficarra answered. "I repeat that she speaks Italian. She spoke only Italian with me. I don't understand a word of English."

I remembered my interrogation, when they yelled that if I didn't stop lying and tell them who had killed Meredith they would lock me up for thirty years. That was still their goal. I was terrified now that I was the only one who saw through them.

The police were not the only people whose testimony damaged me. In one of the great ironies of the trial, Rudy Guede, a convicted murderer, also had power over my life.

The gossip at Capanne was that Guede had found God in prison, and when he walked to the witness stand, looking less cocky and more disheveled than during the pretrial, my hope surged. Maybe he'd been seized by his conscience. I imagined that he'd face Raffaele and me and say straight out that neither of us had participated in the murder. But after Guede was sworn in, he uttered just six words: *"Riservo il diritto di non rispondere"*— "I reserve the right not to respond."

Then he stepped down. He didn't look at me or anyone else as he was led through the double metal doors in the back of the courtroom, flanked by guards just as Raffaele and I always were. He wore an expression of blank indifference.

Guede knew his silence could cost us our freedom. But there was no way to make him tell the truth. People have the right not to incriminate themselves—and in protecting himself, he helped to damn us.

Chapter 26

_____

March–July 2009

After I was accused of murder, people read new meaning into everything about me. A hickey on my neck became a scratch from Meredith in her last, desperate moments. An awkward encounter about a dirty toilet became a murder motive. Male friends I brought home became mysterious lovers of questionable character. Rudy Guede's aside to the guys downstairs about my being cute became proof that he would do anything to earn my attention and approval.

People who never met me were as judgmental as those who had. The runaway media coverage meant that everyone in Perugia— the whole of Italy, really—saw my picture several times a week. They heard the prosecution disparage me just as often.

One of the prosecution's key witnesses said they'd seen Raffaele and me out together on the night of November 1, and another said he saw me on the morning of November 2—at the time we said we were at his apartment. Another of their key witnesses, Kokomani, was sure he'd seen us with Rudy Guede on Halloween, indicating that we knew each other. I was 100 percent positive that he couldn't have.

It wasn't necessary for any of these people to be right. It was enough for them to raise doubts, to make it seem that I was lying. They had to be only marginally convincing.

The thought that these witnesses might wow the jury and judges terrified me.

Marco Quintavalle, a storekeeper who lived near Raffaele's apartment, told the court that he saw a girl waiting for the shop to open at a quarter to eight on the morning of Friday, November 2. "She had a hat and scarf obscuring much of her face but what struck me was how pale she looked and the color of her blue eyes . . . she went to the section at the back of the supermarket on the left, where there are the cleaning products. I can't remember if she bought anything."

But when he saw my picture in the paper a few days later, his memory was precise. "I recognized her as the same girl," he said.

When asked if the girl was in the courtroom, Quintavalle pointed at me. "It's her," he said. "I'm sure of it."

I'd gone to the little store once to pick up milk and cereal. Once. I'd never been in the back, where the cleaning products are apparently shelved.

I don't know how it would feel to be "identified" in court when you're guilty, but hearing those words when I wasn't made me flinch as if I'd been struck. The ache was physical. It took maximum self-control for me to stay seated. Unlike the police, who I thought must be lying to cover their mistakes, I knew the shopkeeper believed what he said. Would the jury?

It came down to who was more credible—him or me?

Under cross-examination, my lawyers asked Quintavalle why he hadn't revealed this information when the police came to question him right after our arrest. He said he hadn't remembered. He hadn't wanted to get involved in the murder case and had come forward only at the urging of a journalist friend in August 2008. I relaxed a little. The jury would see what was true and what wasn't.

The media purposely did not. "A New Hole Appears in Amanda Knox's Alibi" and "Witness Contradicts Amanda Knox's Account." News stories like this infuriated my family and friends.

But strangers, no doubt, would think, *There goes Amanda, lying again.*

———————

Nara Capezzali was a widow in her late sixties who lived in an apartment building behind the parking lot across the street from our villa. She testified that she heard a scream between 11 and 11:30 P.M. "It made my skin crawl, to be honest," she said.

She was certain of the time because she took a nightly diuretic and always woke up around 11 P.M. to use the bathroom.

Before falling back asleep, she said she heard footsteps running up the metal stairs by the parking lot. "At almost the same moment," she heard the crunching of feet on gravel and leaves coming from the direction of our driveway. Never mind that our driveway wasn't gravel; it was mostly dirt.

Meredith's room was on the back of our house, as far as possible from Capezzali's. The defense doubted that anyone could have heard these noises across a busy road and behind closed windows with double panes. But the prosecution clung to Capezzali's account, which was a linchpin used to approximate Meredith's time of death.

One of the few points on which the prosecution and defense agreed was that the police had made an inexcusable blunder shortly after the body was found. They prevented the coroner from taking Meredith's temperature for hours, squandering the best chance to gauge her time of death. The second option—analyzing the contents of Meredith's stomach—was far less reliable. The third—Capezzali's memory—wasn't reliable at all.

There were many bad days during my trial. The worst was the afternoon when evidence was presented to establish the time of Meredith's death. Since the judge had ruled that to protect Meredith's privacy the press and public couldn't see her autopsy photos, he cleared the courtroom of everyone who wasn't directly involved in the trial. Pictures of Meredith's dissected stomach were projected onto a screen like the kind used for home movies. I knew that if I looked, I'd have the same reaction as the juror who bolted for the ladies' room. Even more devastating than the actual image of the stomach was knowing it was my friend's.

Throughout the display, the prosecution delivered a primer on the human digestive system. We learned it takes about two to four hours to digest a meal. Meredith's friends had said that they'd started dinner around 6 P.M. Since the food hadn't yet passed into Meredith's small intestine, my lawyers said she died between 9 and 9:30 P.M.—10 P.M. at the latest.

Any later and her stomach contents would have shown up in her small intestine. According to Meredith's friends, she had gotten home at around 9 P.M. On the only computer the police hadn't fried, Raffaele's laptop, the hard drive showed that we'd finished watching *Amélie* and clicked Stop—the last "human interaction" with the computer—at 9:15 P.M. The tight timing gave us an alibi that even the prosecution didn't try to disprove.

Instead they glossed over these facts and used Capezzali's testimony to determine what time Meredith had died. Based on the scream, they decided that she died at 11:30 P.M. Even though Meredith's digestion indicated an earlier time of death, they were fixated on that scream. Meredith had been murdered by 10 P.M., based on her stomach contents, but the prosecutors invented a scenario in which Meredith was home alone between 9:30 P.M. and 11:30 P.M. According to their argument, the sphincter between the stomach and the small intestine tightens at the moment of trauma, and digestion temporarily stops. Left unanswered was what trauma in that two-hour space interrupted her digestion—the same two hours when the prosecution said she was relaxing on her bed with her shoes off, writing an essay due the next morning. They were ignoring basic human physiology and hanging Meredith's time of death on an older woman's urination habits.

What made their theory even weaker was Capezzali herself. She testified that the morning after she heard the scream, some kids ran by while she was cleaning her apartment and told her a girl in the villa had been killed. Then, at around 11 A.M., when she went out to buy bread, she saw posters with Meredith's face at the newsstand.

The problem: Meredith's body wasn't discovered until after 1 P.M. on November 2. When Mignini asked Capezzali if she might have heard the scream on Halloween and not on November 1, she snapped, "I don't remember these things, these hours, these things. I don't remember them anymore."

I was sure there was no way the jury would put their faith in someone who said she didn't remember.

A ntonio Curatolo was a gray-bearded homeless man who appeared in court as bundled up as if he were on the park bench where he spent most of his time. Like Quintavalle, the storeowner, Curatolo didn't come forward until months after our arrest—and then only at the urging of a local reporter. But the media were billing him as Prosecutor Mignini's *"super testimone"*—"super witness."

I'd been surprised and discouraged when I first heard that a homeless man was claiming he'd seen Raffaele and me at the basketball court in Piazza Grimana on November 1—another story the police had leaked to the media long before the trial began. Impossible claims like this kept popping up out of nowhere, putting me under constant attack. And I didn't feel any better when, during a break, Luciano whispered to me, "He's Mignini's personal 'serial-witness.'"

It turned out that this was the third trial in which Mignini had used Curatolo.

Raffaele and I were, Curatolo said, animatedly talking or arguing with each other and occasionally looking over the fence in the direction of the villa. When was this? From 9:30 P.M. to a little before midnight on November 1, Curatolo answered.

I was surprised by his rambling, and frustrated that the court was giving his testimony credence.

The basketball court was made to order for the prosecution. The most direct walk from Raffaele's apartment to my villa was through Piazza Grimana. It was also the place where Rudy Guede was known to play pick-up games and hang out. It was where I'd once tried to shoot hoops with the guys from downstairs and ended up watching from the sidelines. I hadn't argued with anyone there, and I'd never been back, but what if the jury bought this guy's story?

And why was the prosecution bringing it up? If the story was true, we would have had an alibi. If Curatolo had seen us in the piazza that early, we couldn't have committed the murder between 9:30 P.M. and 10 P.M., when the defense believed Meredith died. And if he'd seen us as late as midnight, we couldn't have made Meredith scream at 11:30 P.M., as Nara Capezzali had reported. His account undermined the prosecution's theory.

That's why Mignini wound back the time that Curatolo had last seen us.

"How did you know what time it was?" an obviously irritated Mignini asked.

"It was shortly before midnight when I left Piazza Grimana to go sleep in the park on the other side of the university," the witness said.

"And you left before midnight?" Mignini pushed.

"Yes, between eleven thirty P.M. and midnight."

"And the last time you saw them was before you left Piazza Grimana?"

"Yes."

"So before eleven thirty P.M.?"

"Yes."

————

Hekuran Kokomani had appeared at our pretrial, where the judge deemed him unreliable. But his testimony was critical to the prosecution. He was the lone person who claimed to have seen Raffaele and me together with Rudy Guede.

Mignini asked Kokomani to point me out in the courtroom. Later, under cross-examination, Carlo asked him how he could be sure it was me he recognized. He'd gotten a good look at my face, Kokomani said, and he remembered me because of the gap between my front teeth. With a look from Carlo, I turned to the court and parted my lips like a child showing newly brushed teeth. Before Carlo could point out that there was no gap, Kokomani muttered in bewilderment, "Oh, she doesn't have it anymore."

One jury member tried to muffle laughter.

Kokomani's testimony was a triumph for us. The prosecution looked inept for putting him on the stand.

My confidence grew as each of Mignini's witnesses delivered testimony full of holes and questionable content. The claims made by his super witnesses strained credulity.

———

I dreaded Patrick Lumumba's testimony for his civil trial. It still gnawed at me that I'd never apologized to him. I was sure the man I'd wrongly named would rail against me. He had told the media that he would never forgive me, he'd lied about firing me, and he had called me "a lion," "a liar," and "a racist." His lawyer, Carlo Pacelli, had called me "Luciferina" and said I had "an angel's face with a demon's soul."

To my enormous surprise, instead of trashing me, Patrick's testimony was full of sadness. He was nine or ten when his politician father was kidnapped, and he never saw him again. "We can't prove he's dead, we can't prove if my father is alive." When Patrick was in jail, he was terrified that history would repeat itself. "I had this feeling that I wasn't going to be able to hold my son again . . . To this day, during the night, I have to go check to see if my son is still there."

He described how difficult it was to reopen his pub after the police had shut it down for three months—and how it ultimately failed.

He was also far more forgiving than I'd expected. I wasn't the best waitress, but I was a fine person, he said.

I can only guess why Patrick had decided to tone down his anti-Amanda commentary. Either he felt he had to be honest under oath or his lawyer had advised him to act meek and likable—and let the venom be rained down by Pacelli himself. Whatever the reason, Patrick told the court, "We always had a good relationship."

———————

Then it was my turn.

At first my lawyers said letting me testify was a risk. I could be provoked. They worried the prosecution would push me to unwittingly say something incriminating. I'd fallen for Mignini's word-twisting when he interrogated me in December of 2007. I'd dissolved into tears at my pretrial.

But I was adamant. "I'm the only one who knows what I went through during the interrogation," I told Luciano and Carlo. "Having you defend me isn't the same as defending myself. I need to show the court what kind of person I am."

I felt it was crucial that I testify. I wanted to talk about my relationship with Meredith. I needed to explain my behavior in the wake of her murder.

Raffaele didn't testify. That may have been the right choice for him. Most of the media attention had landed on me—

Raffaele was seen as someone who had gone along with his evil girlfriend.

In testifying, I wanted to make a point: You guys make me sound like I was crazy that I found three droplets of blood in the bathroom sink and didn't call the police immediately. But I was a twenty-year-old who handled the situation the same way a lot of inexperienced people would have. It's easy to look back and criticize my response, but when I went home that day I didn't know there had been a break-in or a murder. To me, it was a regular day. Yes. The door was open. But I'd known since I moved in that the lock was broken. Maybe it was a cause for concern, but I just figured one of my roommates was taking out the trash or had run to the corner store. I was focused on getting ready for our romantic weekend in Gubbio. My thoughts were mundane. *I'll grab a shower. I'll pack. I'll get back to Raffaele's, and we'll go.*

I knew I wasn't going to convince Mignini that I was innocent. He was too invested in bringing me down to be open to changing his mind. That was okay. I was addressing the people who hadn't yet taken sides.

My lawyers' advice was straightforward: "Remain calm. We'll ask the questions. No one will rush you. It's okay to say, 'I don't know.'"

I didn't want to say, "I don't know," but in the two years since the interrogation, my memory of it had grown hazy. I didn't want to hang on to that night. I'd tried to let it go. I wondered if people would understand how traumatizing it was.

During the weeks leading up to my testimony, I was nervous. What sort of questions would the prosecution ask? Would I be emotionally up to speaking?

But when anyone asked me what I was thinking about my

court dates—June 12 and 13—I'd take a deep breath and say, "I'm ready, I'm ready. It just needs to happen."

I knew Mignini liked to intimidate people. I gave myself a pep talk. *He scared and surprised you the first and second times. But three times? I don't think so!*

As the date got closer, I slept little and talked less. Journalists reported that I was pale and had dark circles under my eyes.

True. I was wearing my anxiety on my face. The day before I had to testify, a nasty cold sore appeared on my lip. My mantra for myself ran through my mind. *You are not afraid. You are not afraid of Mignini. This is your chance.*

When I saw the prosecutor in court, Mignini seemed like a blowhard in a silly robe. I wished I had felt that way when he questioned me before.

The first person to question me was Carlo Pacelli, Patrick's lawyer. Lawyers technically aren't allowed to add their own commentary at this point, only to ask questions. But he made his opinions known through pointed questions like "Did you or did you not accuse Patrick Lumumba of a murder he didn't commit?" and "Didn't the police officers treat you well during your interrogation?"

The lawyer looked disgusted with me. I sat as straight as I could in my chair and pushed my shoulders back—my I-will-not-be-bullied stance.

Within a few minutes I realized that the interpreter hired to translate my English into Italian—the same useless woman I was assigned earlier in the trial—wasn't saying precisely what I was saying.

I've finally gotten a chance to speak! If she gets it wrong, I'll lose my chance forever.

"Your Honor, I'd like to speak in Italian," I said politely. I didn't think about whether it would work or whether it was a good idea. All I could think was, *I have been waiting my turn for nearly two years. This is it!*

At least prison life had been good for my language skills.

I was relieved to be able to speak directly to the jury. The hard part wasn't the Italian; it was being an active listener for hours at a time, making sure I heard the questions correctly and that my questioners didn't push me around.

Pacelli tried to insinuate that I'd come up with Patrick's name on my own in my interrogation. "No," I said. "They put my cell phone in front of me, and said, 'Look, look at the messages. You were going to meet someone.' And when I denied it they called me a 'stupid liar.' From then on I was so scared. They were treating me badly, and I didn't know why.

"It was because the police misunderstood the words 'see you later.' In English, it's not taken literally. It's just another way of saying 'good-bye.' But the police kept asking why I'd made an appointment to meet Patrick. 'Are you covering for Patrick?' they demanded. 'Who's Patrick?'"

We went over how I found the room for rent in the villa, my relationship with Meredith, my history with alcohol and marijuana, and what happened on November 2. The prosecution and the civil parties were confrontational. I was able to respond. It took two exhausting days, and there were a few questions I couldn't answer.

I'd purposely tried to forget the emotional pain of the slap to my head. Other memories had become muddled by time. For instance, I remembered calling my mom only once after Meredith's body was found, but cell phone records indicated that I'd made three calls while Raffaele and I were standing in my driveway.

During my testimony, I was clear. I never stumbled or stalled. I just said, This is what happened. This is what I went through.

I relaxed a little when it was Luciano's turn to question me.

"During the interrogation, there were all these people around me," I said. "In front and behind me, yelling, threatening, and then there was a policewoman behind me who did this."

I slapped my own head to demonstrate.

"One time, two times?" Luciano asked.

"Two times," I said. "The first time I did this."

I dropped my head down as if I'd been struck and opened my mouth wide in surprise.

"Then I turned around toward her and she gave me another."

"So you said what you said, and then you had a crisis of weeping. Then they brought you tea, some coffee, some pastries? When did this happen? If you can be precise," Luciano asked.

"They brought me things only after I made declarations"—depositions—"that Patrick had raped and murdered Meredith, and I had been at the house covering my ears.

"I was there, they were yelling at me, and I only wanted to leave, because I was thinking about my mom, who was arriving soon, and so I said, 'Look, can I please have my phone,' because I wanted to call my mom. They told me no, and then there was this chaos. They yelled at me. They threatened me. It was only after I made declarations that they said, 'No, no, no. Don't worry. We'll protect you. Come on.' That's what happened.

"Before they asked me to make other declarations—I can't say what time it was—but at a certain point I asked, 'Shouldn't I have a lawyer or not?' because I didn't honestly know, because I had seen shows on television that usually when you do these things you have a lawyer, but okay, so should I have one? And at least one of them told me it would be worse for me, because it showed that I didn't want to collaborate with the police. So I said no."

Then it was Mignini's turn. "Why did you say, 'Patrick's name was suggested to me, I was beaten, I was put under pressure?'"

As soon as I started to answer, Mignini interrupted with another question. He'd done the same thing to me during my interrogation at the prison. This time, I wasn't going to let it fluster me. I was going to answer one question at a time. Showing my irritation, I said, "Can I go on?"

I described my November 5 interrogation again. "As the police shouted at me, I squeezed my brain, thinking, 'What have I forgotten? What have I forgotten?' The police were saying, 'Come on, come on, come on. Do you remember? Do you remember? Do you remember?' And then boom on my head." I imitated a slap. "'Remember!' the policewoman shouted. And then boom again. 'Do you remember?'"

When Mignini told me I still hadn't proved that the police had suggested Patrick's name, my lawyers jumped up. The exchange was so heated that Judge Massei asked if I wanted to stop.

I said no.

At the end, the judge asked what I thought of as a few inconsequential questions, such as, Did I turn up the heat when I got to the villa that Friday morning? Did we have heat in the bathroom, or was it cold? Rather, the judge was trying to catch me in an inconsistency. Why would I come home to a cold house when I could have showered at Raffaele's?

Then it was over.

In the past I hadn't been great at standing up for myself. I was proud that this time was different.

When the hearing ended, I got two minutes to talk to my lawyers before the guards led me out of the courtroom. "I was nervous when you first spoke," Luciano admitted, "but by the end I was proud of you."

Carlo said, "Amanda, you nailed it. You came across as a nice, intelligent, sincere girl. You left a good impression."

I took this to mean that I didn't come across as "Foxy Knoxy."

For a while during the trial, the guards would let my parents say hello and good-bye to me in the stairwell just before I left the courthouse for the day. My mom, my dad, Deanna, Aunt Christina, and Uncle Kevin were waiting for me there that day. They hugged me tightly. "We're so proud of you," they said.

I hadn't felt this good since before Meredith was murdered.

After another few days in court, the judge called a two-month summer break.

Chapter 27

September 1–October 9, 2009

The court-ordered summer vacation seemed as endless as the summers of my childhood. But in those days, summer meant freedom; now summer meant confinement. Two wasted months in captivity. At the end I would be no closer to getting out than I was on the first day.

In early September Luciano, Carlo, and another lawyer in Carlo's office, Maria Del Grosso, drove to Capanne to see me.

Carlo leaned across the table in the visitors' room. "Amanda," he said. "They're wrong!"

His customary pessimism had vanished. "There was no blood on the knife," he said. "And there was so little DNA present they didn't have enough to get valid results. We have everything we need to overturn the case!"

I leapt up. Bouncing around on the balls of my feet, I had so much to say that the words tangled in my mouth. "Thank you!" I said. "You did it! Tell me! How did you figure it out?"

The proof, he said, was in the papers.

We'd been asking to see the prosecution's notes and test results since before the pretrial. Only by following in the Polizia

Scientifica's footsteps could we understand how the prosecution's DNA analyst, Patrizia Stefanoni, had come up with her information.

The prosecution was legally required to share the evidence, but even on the pretrial judge's orders they hadn't released all of it.

That had been in September 2008. By then it was July 2009. Ten months had passed. On the day the court recessed for the summer, Judge Massei ordered the prosecution to give us the data.

They still held back some information, but within the papers they did give us, our forensic experts found the prosecution had failed to disclose a fact that should have prevented us from ever being charged. There was no way to tie this knife—and therefore, me—to Meredith's murder. I'd always known that it was impossible for Meredith's DNA to be on the knife, and I'd long known that the prosecution had leaked assumed evidence to the media. Now I knew that these mistakes weren't missteps. Stefanoni and her team had made giant, intentionally misleading leaps, to come up with results designed to confirm our guilt.

I was both ecstatic and furious. Thrilled to know I could reclaim my life. Furious that they would have imprisoned me knowing full well that there was evidence that exonerated me. "How could they have done this?" I asked my lawyers. "Please, please explain that to me."

Carlo, who'd never sugarcoated my situation, said, "These are small-town detectives. They chase after local drug dealers and foreigners without visas. They don't know how to conduct a murder investigation correctly. Plus, they're bullies. To admit fault is to admit that they're not good at their jobs. They suspected you because you behaved differently than the others. They stuck with it because they couldn't afford to be wrong."

And for Mignini, appearing to be right superseded everything else. As I found out that summer, the determined prosecutor had a bizarre past, was being tried for abuse of office, and had a history of coming up with peculiar stories to prove his cases. His own case is currently pending on appeal.

In 2002, on the advice of a psychic, he reopened a decades-old cold case. The Monster of Florence was a serial killer who attacked courting couples in the 1970s and '80s. After murdering them he would take the women's body parts with him. Mignini exhumed the body of Dr. Francesco Narducci after the psychic told him that Narducci, who died in 1985, was the Monster and that he hadn't committed suicide, as had been supposed. Instead, Mignini believed that Narducci had been murdered by members of a satanic sect, who feared the Monster would expose them. He charged twenty people, including government officials, with being members of the same secret sect as the Monster.

Mignini had a habit of taking revenge on anyone who disagreed with him, including politicians, journalists, and officials. His usual tactic was to tap their telephones and sue or jail them. The most famous instance was the arrest of Italian journalist Mario Spezi, and the interrogation of Spezi's American associate Douglas Preston, a writer looking into the Narducci case, who subsequently fled Italy.

In the hour we had each week to discuss my case, my lawyers had never thought there was a reason for us to talk about Mignini's outlandish history. Carlo and Luciano told me only when it became apparent that, for Mignini, winning his case against Raffaele and me was a Hail Mary to save his career and reputation.

"The whole story is insane!" I said. I couldn't take it in. It struck me that I was being tried by a madman who valued his career more than my freedom or the truth about Meredith's murder!

———————

I felt calmer when we returned to the courtroom in mid-September. I imagine Raffaele did, too. The prosecution's forensic documents had answered our lawyers' questions.

Standing confidently in front of the judge, Raffaele's lawyer Giulia Bongiorno made a speech that gave me even more cause for optimism. Keeping the raw data from us until July 30 had violated our rights as defendants. If we'd had it earlier—when we first requested it—it would have altered the trial from the beginning. "The question for the court," Bongiorno said, "is the DNA evidence decisive or not? If you believe it's not, then there hasn't been an injury to the rights of the defense. But if the DNA is decisive, you have to ask yourselves: Did the defense have the possibility to examine the data to be able to counter the conclusions? Did the defense have the diagrams, the electropheragrams, the quantity of DNA, the procedures? You have the answer.

"If you maintain that the missing documents are decisive for the defense, you must nullify the order to stand trial."

Carlo picked up where Bongiorno stopped. "It's evident that the rights of the defense were not fulfilled," he said in a firm manner that presumed the judge and jury would agree. "If I'd had this data, I would have laid out a different defensive strategy from the beginning."

Our lawyers' arguments stirred up all my outrage. The prosecution had kept Raffaele and me in jail for twenty-one months for no reason. If the judges and jury were fair, they'd see that the prosecution had tried to thwart us.

After Carlo and Bongiorno petitioned to end the trial, the prosecution and civil lawyers fired back. They assured the judges and jury that the forensic evidence had been collected and interpreted 100 percent correctly, adding that there was a "mountain of evidence" proving our guilt.

Carlo had cautioned me that Judge Massei would almost certainly not abort the trial. Too much media attention and controversy surrounded it. "That's okay," Carlo said. "Our motion puts the court on notice. They now know that we can and will refute each of the prosecution's arguments."

Even with Carlo's warning that I should not expect a quick end, I let my hopes rise. I'd already spent two frigid winters and two stifling summers behind bars. I'd missed two Christmases and two of my birthdays at home, two years of what should have been carefree college days. I'd missed out on my younger sisters' girlhoods. My mind spun with hopeful possibilities. *What if they lose—today? What if the court accepts our petition to abort the trial? Could this be it? Maybe I'll walk out of Capanne this afternoon!*

That daydream lasted ten minutes. Everyone in the Hall of Frescoes stood for the second time that day. The judge and jury returned to their places. My heart was banging so hard I could hear it pound. *Please, please. Say the trial is over!*

Adjusting his glasses, Judge Massei droned in his unassuming voice, "There will be no annulment. We'll hear both sides discuss the forensic evidence."

I swallowed hard and closed my eyes, willing my tears back in their ducts.

"We're hearing Dr. Gino first today, is it?" Judge Massei asked.

Going or not going to these hearings was the single choice I was allowed to make. I chose to go. I took my seat and started listening.

—————

No one was contesting the brutality of Meredith's death— only how it had happened and who was responsible.

Everyone believed that Rudy Guede had been there and that he had killed Meredith. He was already serving a thirty-year sentence for her sexual assault and murder.

The goal of the prosecution was to prove that I had been there, too.

During the testimony phase, from January to July, witnesses discussed everything from my housekeeping habits to my character and sexual activity. It was intensely personal, and sometimes mortifying.

Picking up after the summer break, the forensics phase lasted only three and a half weeks, but it was still interminable: hour after hour of examination and cross-examination. Witnesses were called to talk about the knife, the bra clasp, my "bloody" footprints, how my DNA could have mixed with Meredith's blood in the bathroom, and our alleged cleanup of the villa. Each expert explained how the evidence was found and documented, how results were calculated and interpreted. They were dissecting a crime I hadn't committed, blaming me using terminology I didn't know. I felt like an observer at someone else's trial. The experts would say things like "Amanda's DNA was on the knife handle," and I would think, *Who is this Amanda?*

I'd rest my chin in my hand, trying to look contemplative—a skill I'd developed during boring college lectures. But no matter how hard I tried to focus, my attention would wander, my head would bob, and the *agente* standing behind me would awaken me to the nightmare. More than feeling embarrassed, I was terrified that my inattention would be interpreted as my not caring and become another mark against me—even though some of the jurors also habitually dozed off.

When testimony wasn't dull, it was disturbing. I couldn't stand thinking about Meredith in the starkly clinical terms the scientists were using to describe her. Did her bruises indicate sexual violence or restraint? What did the wounds to her hands and neck suggest about the dynamics of the aggression? What did the blood splatter and smears on the floor and armoire prove about her position in relation to her attacker or attackers?

The hearings were tedious, gruesome, and enormously upsetting. But we were no longer at the crippling disadvantage we'd been at for two years. Now that the prosecution had been made to show their notes, testing, and some of the raw data, we finally had facts. And the facts supported what I had always known: Raffaele and I had had nothing to do with Meredith's murder. Meredith had never come into contact with Raffaele's kitchen knife. I hadn't walked in her blood.

I wanted to cross-examine the prosecution. I wanted to ask how the story presented with such authority could be so at odds with the truth. Thank God we had Dr. Sarah Gino and Dr. Carlo Torre, both forensics professors at the University of Turin, in addition to Dr. Walter Patumi, out of Perugia. They took the stand and, one by one, began demolishing each of the prosecution's claims.

On the witness stand, Marco Chiacchiera of the Squadra Mobile had explained that "investigative intuition" had led him to the knife. That flimsy explanation did not help me understand how the police could pull a random knife from Raffaele's kitchen drawer and decide that it was, without the smallest doubt, the murder weapon. Or why they never analyzed knives from the villa or Rudy Guede's apartment.

Then we heard the prosecution's hired forensic experts describe the knife as "not incompatible" with Meredith's wounds.

I wasn't the only person who was perplexed. The experts debated the meaning of this phrase as intensely as they did the physical evidence being presented.

During cross-examination, Carlo demanded, "'Not incompatible?' What does that even mean? If the knife was compatible, wouldn't you have written 'compatible'? You wouldn't have bent over backward, twisting words around to create this ambiguous meaning. 'Not incompatible'? Am I to understand, perhaps, that the confiscated knife is 'not incompatible' if only because it's a pointy knife with a single sharpened edge? Am I to understand that *any* pointed knife with a single sharpened edge—*most knives*—would equally qualify as 'not incompatible' with Meredith's wounds? Yes?"

"Yes," the expert answered.

During the afternoon hearing, it turned out that Raffaele's knife was, in fact, *not compatible*. The blade was too wide to have inflicted Meredith's two smaller wounds.

The third and fatal wound was a gash to the throat. The pathologist said Meredith had been stabbed at least three times in the same spot. But the blade of Raffaele's kitchen knife, at 6.89 inches, was longer than the wound was deep—by more than 3.5 inches. Under Carlo's questioning, Professor Torre, a serious man in his sixties who favored lime-green glasses, explained that in a moment of homicidal frenzy, it would be highly unlikely for a killer to plunge a knife in only halfway, to 3.149 inches. And the odds would rise to impossible when you considered driving a knife in, to precisely the same depth, measurable to a thousandth of an inch, three times in a row. Torre brought in a foam bust and an exact copy of the knife to demonstrate how implausible this feat would be. I thought it was a good idea, but I couldn't watch anyone stab anything—even a dummy. The notion that anyone thought I could have done that to a person—to my friend—made me not just heartsick but feeling like I might throw up. I squeezed my eyes shut.

As he had done when the prosecution showed Meredith's stomach contents, the judge cleared the press and public from the courtroom so photos of Meredith's wounds could be projected on a pull-down screen. These were the same deeply disturbing autopsy photos Carlo had tried to show me seventeen months before. I knew then that I could never stand to look at them.

It kept me from glancing up.

But I couldn't choose not to hear. Dr. Torre said there was a scratch at the top of the lethal wound. The pressure had been just enough to nick the skin, he said, adding that the scrape was made from the hilt. The only way this could have happened was if the full length of the blade penetrated Meredith's neck. More proof that Raffaele's knife could not be the murder weapon.

At the next hearing Manuela Comodi, the co-prosecutor in charge of forensics for the trial, swept into the courtroom triumphantly carrying a flat cardboard box, a little smaller than the ones used for carryout pizza. After opening it, Comodi paraded it in front of the court, as though she were displaying the queen's jewels. Her pride showed on her face as the jurors and experts stood up, straining in her direction to get a good look at what was inside—the knife that had been confiscated from Raffaele's apartment was wrapped in a baggie. Only Comodi was allowed to touch it, to pick it up and hold its plastic-shrouded blade up to the light.

Her theatrics were exasperating. The prosecution continued to say that it was Meredith's DNA that had lodged in a small scratch on the knife blade. The prosecution still claimed this as incontrovertible proof that I had used the knife to kill Meredith. I knew it to be a regular kitchen knife that had last been used to prepare a salad.

After everyone had had a good look, Comodi gingerly closed the box and left the courtroom.

———————

A DNA reading is a series of peaks that looks like an EKG. By analyzing the size of the peak in thirteen or more locations, scientists can be almost certain they have a DNA profile unique to one person—or that person's identical twin. Done correctly, the reading is more accurate than a fingerprint.

During the pretrial, Stefanoni testified that she had tested enough DNA from the knife to get an accurate reading. But now, a year later, Dr. Gino had seen the raw data, including the amount of DNA that was tested. If there was any DNA there at all, it was too little to determine using the lab's sensitive instruments, Gino said. Stefanoni had met none of the internationally accepted methods for identifying DNA. When the test results are too low to be read clearly, the protocol is to run a second test. This was impossible to do, because all the genetic material had been used up in the first test. Moreover, there was an extremely high likelihood of contamination in the lab, where billions of Meredith's DNA strands were present.

The prosecution said Stefanoni's methods were perfectly acceptable, because in proving that the knife was the murder weapon, she'd "struck gold." But an unbiased analyst would have thrown out the results.

What I couldn't understand was why this infinitesimal, unconfirmed sample found on a random knife that didn't correspond with Meredith's wounds or the bloodstain on the bedsheet—the murderer's signature— held any sway. Copious amounts of Rudy Guede's genetic material had been found in Meredith's bedroom, on her body, in her purse, and in the toilet.

Perhaps most telling, when the knife was tested for blood, not even a diluted trace was found, evidence that should have convinced the prosecution's scientists that any DNA that might have been found there had come from contamination—not from a cut.

But the prosecution didn't admit they had tested for blood until our experts found out for themselves.

The situation was similar to the prosecution's claim throughout the investigation, the pretrial, and now the trial that my feet were "dripping with Meredith's blood." My lawyers and I had spent hours trying to figure out why they thought this. We knew that investigators had uncovered otherwise invisible prints with luminol. Familiar to watchers of *CSI*, the spray glows blue when exposed to hemoglobin. But blood is not the only substance that sets off a luminol reaction. Cleaning agents, bleach, human waste, urine stains, and even rust do the same. Forensic scientists therefore use a separate "confirmatory" test that detects only human blood, to be sure a stain contains blood. Had the Polizia Scientifica done this follow-up test?

Under cross-examination during the pretrial, Stefanoni was emphatic. "No," she responded.

It wasn't until Dr. Gino read the documents Judge Massei had ordered the prosecution to share with us that she, and then the rest of my defense team, began seeing a pattern. As with the knife, it turned out that Stefanoni's forensics team had done the TMB test and it came out negative. There were footprints. But they could have come from anything—and at any time, not necessarily after the murder. What matters is that there was no blood.

On the stand, Stefanoni declared that the negative blood test was irrelevant. We knew we were looking at blood, she explained, because the luminol glowed more brightly.

"Is it true that luminol glows more when sprayed on blood?" Carlo asked Dr. Gino.

"No."

The prosecution had an answer for everything, even when it meant lying to cover up other lies.

Stefanoni assumed that because both Meredith's and my DNA were found in the hall outside the bathroom, I was connected to the murder. It was a startling mistake for a forensic scientist to make.

Human beings shed thousands of skin cells every hour and nearly a million a day. We all leave DNA wherever we go—when we rest an arm on a counter, eat a spoonful of ice cream, grab a steering wheel, or walk barefoot, as I did when I came home for a shower on the morning of November 2. Of course my DNA would be mingled with Meredith's in the common hallway between our bedrooms—we'd lived in the same house and walked on the same floor tiles for six weeks.

The prosecution had no evidence against us, and worse yet, they'd withheld information likely to prove our innocence.

More infuriating was that Stefanoni continued to argue the prosecution's inaccurate points during cross-examination.

Some things could not be proven or disproven. DNA doesn't show its age. Science has no way of knowing when I left footprints in the hallway or what time I was in the bathroom. Or how long Raffaele's DNA had been on Meredith's bra clasp— the only evidence that tied Raffaele to Meredith's bedroom. It meant that both Raffaele and I were in the same excruciatingly frustrating position.

When the white-suited Polizia Scientifica first swept the crime scene on November 2 and 3, the little strip of fabric with the bra fastener lay under the bloody cushion beneath Meredith's body, cut from the rest of the bra. The forensics team put a placard beside it, assigning it the letter *Y*. But when they bagged the evidence in Meredith's bedroom and sent it to the forensics lab where Stefanoni worked in Rome, sample *Y* was overlooked and left behind.

Six weeks later, when the Polizia Scientifica returned to No. 7, Via della Pergola, they spotted the bra clasp again. Only this time, it was a yard from where it started, lying beneath a rolled-up carpet and a sock. Between the forensics team's two trips, other police units had ransacked the villa. Unlike the Polizia Scientifica, those units had made no pretense of keeping the crime scene safe from contamination.

Replaying the video of this second trip into the villa, Raffaele's forensics expert pointed out, "The clasp goes from one scientist to another, and we don't see gloves being changed. We then see it being put on the floor and picked up again. These procedures are all wrong . . . By not changing gloves and by touching other objects, cross-contamination of DNA is highly possible."

When Stefanoni was asked how the bra clasp got from one spot to another without being contaminated, she responded, "*È traslato*"—"It moved"—the same phrase Italians use when they're talking about religious miracles.

"It didn't get contaminated in that process?" Raffaele's DNA expert asked her.

"No."

"Why?"

"Because DNA doesn't fly," she snapped.

"It was trampled and dragged across the floor and you're saying there's no possibility it was contaminated?"

Had Raffaele been in the room, his DNA would have been as abundant as Guede's. It would be illogical to suggest that it was left on a single small hook on Meredith's bra and nowhere else. Furthermore, one of Raffaele's defense experts pointed out that the genetic profile was incomplete, and could have matched hundreds of people in Perugia's small population. But the main point is that this piece of cloth and metal had been underfoot and moved around by the dozens of people who went through the house in the six weeks since it had first been photographed. The contents of the room had been moved, and many items piled in heaps. The cloth fragment had clearly been moved around the floor, and who knows where else.

The prosecution worked hard to convince the judges and jury that their forensic findings made sense.

One morning, Manuela Comodi, the co-prosecutor, told the

court that to show her dedication to the case, she had brought in her own bra.

She was carrying a white cotton underwire bra, the closest match in her drawer to what Meredith had been wearing, although, she said, chuckling, it was larger than Meredith's. Comodi hung the bra on a hanger to mimic a person wearing it. Using her index finger, she showed the mesmerized court how Raffaele could have hooked his finger to pull the back strap of Meredith's bra (somehow leaving DNA on the clasp but not the cloth) and then sliced off the fastener section with a knife.

Jury members tittered. The explanation and demonstration were absurd, but no one looked skeptical. *Are people actually buying this?*

Another day, the prosecution said that finding my DNA in the bathroom was proof I'd been involved in the murder. They didn't consider that I had lived in the villa and used that bathroom every day for weeks. Even rookie forensic scientists know that roommates leave DNA in bathrooms, but the prosecution insisted it was incriminating evidence. They claimed that the only way my DNA could have been collected with the samples of Meredith's blood was if I'd been washing her blood off my hands.

The prosecution said they were certain the murder had been a group attack. Why, then, was none of my DNA or Raffaele's DNA in Meredith's bedroom? Their answer: because Raffaele and I had scrubbed the crime scene clean of our DNA, leaving only Guede's.

That theory gave me super powers. DNA is not something you can cherry-pick; it's invisible. Even if I could somehow magically see DNA, there is no way I could tell one person's DNA from another's just by looking—no one can.

The prosecution contended that, as representatives of the state, they were the impartial party and maintained that their conclusions were legitimate. Our experts, they said, couldn't be trusted because they were being paid to defend us. And our critiques, objections, and conclusions were just smoke screens created to confuse the judges and jury.

The divide between experts for the defense and the prosecution grew wide and bitter. The two sides had reached a stalemate. Both defense teams decided that an independent review of the evidence was essential.

I'm sure some people thought we were grandstanding when Carlo asked the judge to order such a review. I didn't want to extend my time in the courtroom, or to make the Kerchers sit there an extra minute. It distressed me that Meredith's family thought I was guilty, but I always had huge empathy for them. No matter the verdict, they would leave Perugia without their daughter and sister. I knew their pain would stay fresh, casting sadness over everything good. But I also knew I had to ask for an independent review. What was at stake for me was how I would be allowed to live my life. No one else's future depended on this trial except Raffaele's and mine.

The court's deliberation over whether to grant an independent review was unnervingly quick. I sat between my lawyers for just fifteen minutes. "What do you think they'll decide?" I asked Carlo and Luciano. "It's hard to see why they wouldn't grant it. It's the only fair thing to do."

"We made a legitimate argument," Carlo answered, "but it's hard to tell with this judge."

"It's not the end for us if the request isn't accepted," Luciano said. "It doesn't mean we've lost. *Coraggio*—courage—Amanda."

When the court came back in, I squeezed Luciano's hand under the table and waited, barely able to breathe.

With zero fanfare—the way he did everything—Judge Massei stood at the microphone and announced, "There will be no independent review. The court has heard enough expert opinion to make a decision in the case."

This was by far the biggest blow yet. Carlo and Luciano looked weary and disappointed, and neither met my eyes that afternoon.

But I was still so blinded by hope, and my faith in my own innocence, that I actually read this news as positive. I could be accused, but they couldn't possibly convict me of something I hadn't done. There was only one honest outcome. I couldn't imagine that the jurors would side with the police without question. They couldn't ignore everything that our defense had put forth. "They must think we don't need the review because there's already enough reasonable doubt," I said to Luciano.

He patted me on the arm but didn't answer.

I was convinced that my perspective was right. *If the two sides are saying completely different things, that has to mean reasonable doubt. I'll take reasonable doubt. That's good enough for me.* I really did feel that turn meant that my freedom was near.

After so many witnesses, and so many words over so many months, there were no more questions to be asked or answered. The judge announced that the court would adjourn until November 20, to allow the prosecution and defense lawyers to prepare their closing arguments. I couldn't believe I had to wait six weeks! Barring an emergency, with the finality of a curtain drop, the court would render a verdict on Friday, December 4.

Chapter 28

October 10–December 4, 2009

I n the weeks leading up to the closing arguments, I put our chances of winning at 95 percent.

Carlo gave us fifty-fifty. "Judge Massei challenges the defense a lot more than he does the prosecution," he said. "And the judges and jury nod whenever the prosecution or the Kerchers' lawyer talks, but look bored when it's our turn."

Still, I held tight to optimism.

Not without reason. Journalists told Mom and Dad they weren't convinced by the prosecution's arguments. Even the Italian media, uniformly negative since the beginning, seemed to be turning around. A show I saw on the second anniversary of Meredith's death replayed Rudy Guede's first recorded conversation, in which he said that I wasn't at the villa. *If the press can see the truth, surely the judge and jury can, too.*

I got daily mail from strangers who had faith in me. And now that the forensic information was public, two renowned DNA scientists, Dr. Elizabeth Johnson of California and Dr. Greg Hampikian, a professor and head of the Idaho Innocence Project, had written a letter of concern signed by seven other experts from around the United States. "No credible scientific evidence has been presented to associate this kitchen knife with the murder of Meredith Kercher," the report read. The problem with the bra clasp, it said, was contamination. The scientists concluded that the DNA evidence on the knife and the bra clasp "could have been obtained if no crime had occurred."

The science would win the day. I would be acquitted, if not outright, then for reasonable doubt. The prosecution's talk was just that—talk. It wasn't enough for an intelligent jury to convict me.

But sometimes my confidence flagged, and I felt a sickening feeling in the pit of my stomach. In high school I'd learned that 95 percent of criminal cases in the United States end in conviction, and I couldn't get that statistic out of my head. I was too afraid to ask if it was the same in Italy.

The prosecution had undermined my credibility in every way possible. Mignini had called my family "a clan"—intending a Mafia connotation—that falsely proclaimed my innocence. He and his co-prosecutor, Manuela Comodi, had argued that my lawyers' pitch for reasonable doubt was a technique to create confusion and prolong the Kerchers' grief.

A public opinion poll on TV said that more than 60 percent of Italians thought I was guilty. The people who only watched television reports most likely sided with the prosecution. That realization spawned a deep-down fear that I'd be convicted, my innocence be damned.

Prisoners gossiped about my case all the time, behind my back and to me. "Come on, Amanda. You can tell *me*."

Over the many months of the trial, I grew more and more numb, too afraid to feel lest my emotions overwhelm me. But the closer we got to a verdict, the more my anxiety spiked. I started losing my hair. Each time I washed it, a huge clump would come out in my hands. I cried suddenly over nothing. Panic attacks left me gasping for air. My energy was so low that just walking made me stiff and dizzy. Desperate to shut everything out, I climbed into bed each night at seven or eight, earplugs in, covers over my head.

Guards often sent me to see Don Saulo. He was the only person who calmed me. I'd sit quietly on his couch holding his hand. One of the few comments he made was "I hope you will go home. As far as I know, you are innocent, and you don't belong in prison."

That meant the world to me.

Another day, Don Saulo suggested I try praying. "You can just say, 'God, if you exist, I really need your help right now.'"

So I did pray. It made me feel ridiculous, since I don't believe in God. But I was also relieved. I thought, *I'm covering my bases, just in case.* My conversations with God were always the same: "Look, I know innocent people suffer—Meredith didn't deserve to die. But I don't think I can handle this. Please let it be part of your plan not to have me go through this, because I don't know how."

Deep down, I didn't believe a bad outcome was possible. My confidence always overrode the worst-case scenario. My mind *could not* go there.

Instead, I daydreamed about home.

I thought, *In my new life, I want to be healthy and productive and musical.*

I cut out décor ideas I liked from magazines and sent them to Madison, my friend from college, for our new apartment in Seattle. I worried about finding a job. I didn't want to have to rely on my parents. I'd already cost them way too much time and money and caused too much emotional upheaval.

When they visited me at Capanne, my family would ask, "What's the first thing you want to do when you get home? What if we escape to Arizona together"—where Chris is from—"to go rock climbing? We can go where no one would find you."

My wants were simpler. I imagined being surrounded in the doorway of my mom's house in West Seattle by everyone who meant something to me—my family; friends like Madison, DJ, James, Andrew, and Brett; my soccer team girlfriends; teachers. All the people who gathered there for the once-a-week, middle-of-the-night ten-minute phone call I'd been allowed to make home for the past two years.

If the trial goes badly, I thought, *I'll cut my hair.*

It was a superficial, stupid, melodramatic idea. But it was as far as I could go to wrap my mind around an ending that was too enormous and too terrifying to handle.

One day I got up the courage to ask Chris, who was in Perugia leading up to the closing arguments and verdict, "What would a conviction mean?"

So afraid to acknowledge that uncharted, dark place, I could only whisper.

"There would be an appeal, and if you didn't get acquitted, then the Supreme Court would exonerate you. At the most, Amanda, it would take five years," Chris explained.

"Five years?!"

That was way more than I wanted to know.

Chris jumped in to reassure me. "If that happened, Amanda, we'd find a way to save you! But don't worry! It's *not* going to happen! And if for some utterly bizarre reason it goes the wrong way, I'm moving to Italy."

Chris was already doing his IT job from the cold, stark apartment my parents had rented on the outskirts of Perugia, but his promise sounded so drastic it underscored the absurdity of a conviction.

There was only one person who, while she cheered me on, also cautioned me against what she called my "Mickey Mouse view." An inmate in her mid-fifties, Laura was my new ally in prison. She'd been transferred to Capanne over the summer. As the only two Americans there, we'd bonded immediately over how displaced we felt. Leading up to my verdict, she often said, "Amanda, you have the optimism of a Disney movie, but that's not how the real world is. Things don't always work out just because they should."

A nd then, just like that, the final act was upon us.
    We'd been going to court, trying our case once or twice a week, for so long that it seemed we'd been living in a suspended state forever. But as we neared the end, it was like somebody had pushed the Fast Forward button. Hearings were now held nearly every day, one tumbling into the next, all leading toward the verdict, scheduled for the first Friday in December. My family bought me a plane ticket home to Seattle as soon as they found out when the verdict was due. "We're going to get you out of here and back home," they promised.

Prosecutor Giuliano Mignini gave his closing argument first. Alternating between a calm, almost quiet recitation of the "facts" and the fiery rants of a preacher at a tent revival, Mignini summarized Raffaele's and my part in the savagery that took Meredith's life. He started with the idea that Filomena's window was too high to be a credible entry point into the villa and ended with our tossing Meredith's stolen British and Italian cell phones over the garden wall.

Raffaele and I had accused "this poor Rudy," as Mignini called him, of "being the only one" to attack Meredith. "He has his own grave responsibility, but the responsibility is not only his own," Mignini intoned.

I couldn't believe what the prosecutor was saying. He, who was championing himself as the bearer of truth for Meredith's family, was calling the murderer "Poor Rudy"? Evidence of Rudy's crimes was everywhere, and his history of theft matched the burglary. Poor Rudy? Guede had stolen! He had killed Meredith! He had left a handprint in Meredith's blood! He had fled! He had lied! Poor Rudy?

Mignini went on. "For Amanda, the moment had come to take revenge on that simpering girl, that's how she thought, and in a crescendo of threats and violence, which grew and grew, the siege on Mez"—Meredith's nickname—"began."

Revenge for what?!

"By now it was an unstoppable game of violence and sex. The aggressors initially threatened her and demanded her submission to the hard-core sex game. It's easy to imagine Amanda, angry at the British girl for her increasing criticism of Amanda's sexual easiness, reproaching Mez for her reserve. Let's try to imagine— she insulted her. Perhaps she said, 'You were a little saint. Now we'll show you. Now you have no choice but to have sex.'"

He's perverse! How did he come up with such a twisted scenario? He's portraying me as a psychopath! Is Mignini allowed to put words in my mouth, thoughts in my head? I would never force anyone to have sex. I would never threaten or ridicule anyone.

Mignini continued: "The British girl was still on her knees with her head turned toward the armoire. Rudy was to the left of Mez. Raffaele brought himself around behind her and tried to tear off the infamous bra clasp and there was the successful cutting of it. Amanda was in front of Mez with her back to the armoire, wielding the knife from below, pointing it upward toward Meredith's neck.

"Raffaele also took out his knife," Mignini said. "And used it to threaten and wound Mez from the right."

I went from seething to stunned. For the entire eleven-month trial there was only one knife discussed. Having heard the evidence proving that the kitchen knife was not the murder weapon, Mignini had now invented a second knife—a knife that has never been found or even mentioned. In fact, the police had confiscated Raffaele's entire pocketknife collection. None had shown any trace of blood or of Meredith's DNA, and the double-edged blades couldn't have made Meredith's wounds. But suddenly Mignini was saying that Raffaele had used another knife he'd somehow stashed away. I knew why. Not only did Mignini's fantastic scenario explain away the bloody imprint on Meredith's bedsheet and the wounds that couldn't have been made with the kitchen knife, but it also allowed the prosecutor to put a knife in Raffaele's hands. Otherwise, Raffaele had no role in the murder.

Mignini kept going: "By now, seeing the resistance of the victim and the growing rage of the aggressors, who realized the British girl wouldn't give in, she wouldn't submit herself to rape, the game had to end. Amanda provoked the wound on the right side of Mez's neck and tried to strangle her friend . . . It is a plausible reconstruction. Obviously it is a hypothesis . . . At this point it's probable that Mez, realizing that the violence was unstoppable, made that terrible and desperate scream . . . Amanda provoked the deepest wound, the one on the left . . . Mez collapsed onto her right side. One of the aggressors looked for the girl's cell phones, probably Raffaele, and he set down one of the knives on the bedsheet."

I hope people are hearing Mignini saying, "it's probable" and "it's a hypothesis." You can't convict someone based on a hypothesis that the evidence doesn't support!

As for my interrogation at the *questura*, Mignini described the interpreter—the woman who had called me "a stupid liar" and had told me to "stop lying"—as "very sweet." "I remember that evening how she behaved toward Amanda," he said.

Then he recalled from earlier in the trial, when Judge Massei questioned me about my interrogation. "Your Honor asked, 'But a suggestion in what sense? Did they tell you, 'Say that it was Lumumba?' Because a suggestion is just that . . . And Amanda said, 'No. They didn't tell me that it was him.' And so what suggestion is it?

"Amanda said, 'But they told me, Ah, but we know that you were with him, that you met with him.' The police were doing their job . . . they were trying to make this person talk . . . These are the pressures, then. Completely normal and necessary investigative activity. There were no suggestions because a suggestion is: Say it was Lumumba."

Mignini knew how my interrogation had gone. The police were yelling that I knew who the murderer was, that I had to remember, that I'd gone out to meet Patrick that night. They made me believe I had trauma-induced amnesia. They threatened me if I didn't name the murderer—even though I said I didn't know who the murderer was! *How is that* not *suggestion? How is that* not *coercion?*

Mignini's rant lasted one day, from 9 A.M. to 4 P.M. When I went back to Capanne that afternoon, I felt as though I'd been beaten with a hammer. But I had survived. I repeated the child-hood adage my mom used to recite: "Words can never hurt me." I wished that were true.

The co-prosecutor, Manuela Comodi, spoke the next day. She talked about the forensic evidence as if each element were a neatly laid brick and they all fit together. She had built a brick house that she contended proved we were guilty. We heard for the umpteenth time that Stefanoni was right and our forensic experts were biased. Then she introduced a new motive—a nonmotive, actually—a one-size-fits-all, everybody's-doing-it explanation for anyone who questioned Mignini's revolving mo-tives. Squaring her shoulders, she told the jury, "We live at a time where violence is purposeless."

Echoing Mignini, Comodi said it was reasonable to assume that Raffaele had brought along a pocketknife, which he used to poke Meredith in the neck to scare her.

Then the prosecution turned off the lights.

"I'd like to show the court a visual prop we've constructed to demonstrate our theory of the murder," Comodi said.

This introduced the most surreal moment of my nightmarish trial: a 3-D computer-generated animation with avatars repre-senting me, Raffaele, Rudy Guede, and Meredith.

Carlo and Luciano were apoplectic. They shouted their objec-tions, insisting that the film was unnecessary and inflammatory.

Judge Massei allowed it. I didn't watch it, but my lawyers said the avatar of me was dressed in a striped shirt like one I often wore to court. Raffaele, Guede, and I were depicted sneering. Meredith's avatar had an expression of horror and pain. The car-toon used real crime scene photos to show the blood splatters in Meredith's room.

The animation dramatized the prosecution's hypothesis, showing Raffaele and me leaving his apartment and sitting at the basketball court in Piazza Grimana, me arguing with Meredith at the house, the three of us attacking her.

I kept my head down, my eyes on the table. My stomach was churning. The courtroom was suddenly hot. I was boiling with anger and near tears. *How are they allowed to make up what happened?* I tried to block out Comodi's voice as she narrated the imagined event.

The cartoon couldn't be entered as evidence, so no one outside the courtroom saw it. But the prosecution had achieved their goal. They'd planted an image in the minds of the judges and jury.

When the lights came up, Comodi closed with a straightforward request: Give Amanda and Raffaele life imprisonment.

After Comodi came Patrick's civil attorney, Carlo Pacelli. Unlike Patrick, whose testimony had been fair, Pacelli trashed me mercilessly.

"Who is Amanda Knox? The Knox who is unscrupulous in lying, in slandering; beautiful, intelligent, cunning, and crafty is above all how she appears before you, and how she appeared before you during more than forty hearings: very feminine, cute, enchanting, a white face, blue eyes, simple, sweet, naïve, fresh-faced, with a family at her back and parents who, even if separated, are loving and affectionate.

"Is Amanda Knox the daughter who everyone would want? The friend who everyone would like to meet? Yes. Great. The defense counselor says that Amanda is exactly as you see her today, in this courtroom, as she appears. She's exactly this. But the defendant that you see, Your Honors, is a student transformed by a long prison detention . . . And so the question that arises . . . who was Amanda Knox on the first of November?"

Then he descended on me as if I were a witch on trial in the Middle Ages.

"So who is Amanda Knox? In my opinion, within her resides a double soul—the angelic and compassionate, gentle and naïve one, of Saint Maria Goretti, and the satanic, diabolic Luciferina, who was brought to engage in extreme, borderline acts and to adopt dissolute behavior. This last was the Amanda of November 1, 2007 . . . It must be spelled out clearly: Amanda was a girl who was clean on the outside because she was dirty within, spirit and soul . . ."

Thank God Italy doesn't believe in burning people at the stake anymore! Pacelli is piggybacking off the prosecution's baseless accusations! How can he live with himself? How can any of them?

The Kerchers' civil attorney, Francesco Maresca, emphasized the horror that had been inflicted on Meredith—by a group. He knew this because had it been only one attacker, there's no doubt that Meredith, who knew karate, would have defended herself.

How can any girl defend herself against a guy armed with a knife?

"It's a very long list of lesions: to the face, neck, hands, forearms, thighs. Try to understand the terror, the fear, the pain this girl suffered in the last seconds of her life in the face of the multiple aggression, an aggression brought about by more than one person."

Maresca didn't mention that the prosecution's own coroner—the only person who'd analyzed Meredith's body—had said it was impossible to determine whether one or more people attacked Meredith.

Maresca, like Mignini, criticized any media that had questioned his work. But what most enraged me was the false contrast he set up between the Kerchers and my family.

"You'll remember Meredith's family for their absolute composure. They taught the world the elegance of silence. We've never heard them on the television . . . in the newspapers. They've never given an interview. There's an abysmal difference between them and what has been defined as the Knox Clan and the Sollecito Clan, which give interviews on national television and in magazines every day."

Thank God for my "clan," I thought. *They're the only ones on my side.*

It was wrong of Maresca to compare my family to Meredith's. I knew that the Kerchers were loving parents and good people because of the way Meredith had talked about them. She knew the same about mine. One of the things that connected us was that we were both close to our families. *Meredith's family is grieving, but my family knows that I'm not the cause of the Kerchers' grief. Just as Meredith's family came to Perugia to seek justice for their daughter, mine have come to seek justice for me. Both families are good. Both families are doing the best they can, the best way they know how.*

———

Finally it was our turn.

Thank God we've arrived in friendly territory! I thought.

I was fed up with being the target. Now I was bound by anxiety. The end was so close! Home was on the horizon.

Raffaele's lawyers, Luca Maori and Giulia Bongiorno, worked to put distance between their client and Guede.

"Raffaele and Rudy Guede never met, went out together, or saw each other," Maori said. "The two young men belonged to completely different worlds and cultures. Raffaele comes from a big and healthy family. Rudy rejected his family. Raffaele has always been a model student. Rudy was never interested in school or work. Raffaele is timid and reserved. Rudy is uninhibited, arrogant, extroverted."

"Accomplices who don't know each other . . ." Bongiorno said, drawing out the words to emphasize the paradox that they couldn't have been accomplices if they didn't even know each other!

Raffaele, she told the court, was "Mr. Nobody"—put in by the prosecution as an afterthought. "There was no evidence of him at the scene." The prosecution had contradicted themselves. "He's there, but he's not. He has a knife, but he doesn't. He's passive, he's active."

In defending Raffaele, she also defended me. "If the court doesn't mind, and Amanda doesn't mind, the innocence of my client depends on Amanda Knox," she said. "A lot of people think that she doesn't make sense. But Amanda just sees things her way. She reacts differently. She's not a classic Italian woman. She has a naïve perspective of life, or did when the events occurred. But just because she acted differently from other people doesn't mean she killed someone. . . .

"Amanda looked at the world with the eyes of Amélie" she said, referring to the quirky waif in the movie that Raffaele and I watched the night of Meredith's murder.

Amélie and I had traits in common, Bongiorno said. "The extravagant, bizarre personality, full of imagination. If there's a personality who does cartwheels and who confesses something she imagined, it's her. I believe that what happened is easy to guess. Amanda, being a little bizarre and naïve, when she went into the *questura*, was truly trying to help the police and she was told, 'Amanda, imagine. Help us, Amanda. Amanda, reconstruct it. Amanda, find the solution. Amanda, try.' She tried to do so, she tried to help, because she wanted to help the police, because Amanda is precisely the Amélie of Seattle."

Then, the moment that Luciano, Carlo, Maria Del Grosso (Carlo's second), and I had been waiting for. Just as they'd been promising me for more than two years, they went over the entire case—the witnesses, the forensics, the illogic of the prosecution's case—turning the clock back to the beginning and telling it from our perspective.

"At lunch hour on November 2, 2007, a body was discovered," Luciano began. "It was a disturbing fact that captured the hearts of everyone. Naturally there were those who investigated. Naturally there were testimonies. Naturally there was the initial investigative activity. Immediately, immediately, especially Amanda, but also Raffaele, were suspected, investigated, and heard for four days following the discovery of the body. There was demand for haste. There was demand for efficiency. There was demand.

"Such demand and such haste led to the wrongful arrest of Patrick Lumumba—a grave mistake."

Carlo picked up the thread. "There is a responsible party for this and it's not Amanda Knox. Lumumba's arrest was not executed by Amanda Knox. She gave information, false information. Now we know. But you couldn't give credit to what Amanda said in that way, in that moment and in that way. A general principle for operating under such circumstances is maximum caution. In that awkward situation there was instead the maximum haste."

Having heard what they wanted to hear and without checking further, the investigators and Prosecutor Mignini arrested Patrick—bringing him in "like a sack of potatoes," Luciano said.

I was relieved to hear someone telling the truth. Seeing my lawyers in this theatrical mode, I relaxed the tiniest bit.

Maria Del Grosso criticized Mignini for the fiction he'd invented. "What must be judged today is whether this girl committed murder by brutal means. To sustain this accusation you need very strong elements, and what element does the prosecution bring us? The flushing of the toilet. Amanda was an adulterer. I hope that not even Prosecutor Mignini believes in the improbable, unrealistic, imaginary contrast of the two figures of Amanda and Meredith."

Yes. Make them stop pitting Meredith and me against each other! We were never like that in real life!

"In chambers you will have to apply the law, but remember: condemning two innocents will not restore justice to poor Meredith's memory, nor to her family. There's only one thing to do in this case: acquit."

During the rebuttals, on December 3, each lawyer was given a half hour to counter the closing arguments made over the past two weeks. Speaking for me, Maria criticized Mignini for portraying Meredith as a saint and me as a devil. In reality, she said, we lived similar lives. Meredith had casual sexual relationships. So did I. Meredith wanted to study seriously and be responsible. So did I.

Mignini continued to insinuate that I had loose morals, going beyond the testimony to come up with his own examples. In an eleventh-hour swipe at my reputation, he said it was likely that I had met up with Rudy and made a date with him for the one hour Raffaele had planned to take his friend, Jovanna Popovic, to the bus station the night of November 1. I wanted to amuse myself with another boy—a "not unwelcome distraction."

"She was a little, let's say, very social, Amanda. Amanda was sick of the reproaches of Meredith, who also talked about needing to be faithful to one's own boyfriend, no doubt! Meredith was precisely of an uncommon level of uprightness."

Mignini knows neither Meredith nor me in the least.

"I've asked myself if we were listening to a prosecutor, a lawyer, or a moralist," Maria said, standing up for women everywhere. "Who are you to make such a claim in the name of a woman that it's so much like a woman to be at the throat of another woman?"

Then Raffaele and I made our final pleas. Raffaele talked about how he would never hurt anyone. That he had no reason to. That he wouldn't have done something just because I'd told him to.

I'd spent hours sitting on my bed making notes about what I wanted to say, but as soon as I stood up, every word emptied from my brain. I had to go with what came to me, on the few notes I had prepared.

"People have asked me this question: how are you able to remain calm? First of all, I'm not calm. I'm scared to lose myself. I'm scared to be defined as what I am not and by acts that don't belong to me. I'm afraid to have the mask of a murderer forced on my skin.

"I feel more connected to you, more vulnerable before you, but also trusting and sure in my conscience. For this I thank you . . . I thank the prosecution because they are trying to do their job, even if they don't understand, even if they are not able to understand, because they are trying to bring justice to an act that tore a person from this world. So I thank them for what they do . . . It is up to you now. So I thank you."

My words were so inadequate. But at least I remembered to thank the court again. Now I had to put my faith in what my lawyers and our experts and I had said month after month. I had to believe that it was good enough.

When I went back to prison that afternoon, I saw Don Saulo.

"I'm feeling hopeful," I said. "I think everything is going to work out well. Things have turned around. It's clear the evidence against me is unreliable. There are lots of people who support me. So why do I feel like I'm about to be executed?"

———————

On the final morning, I was glad for the thirty-minute van trip from the prison to the downtown courthouse. It gave me something to do. And even though I'd be leaving prison as soon as the verdict was rendered, I was happy I could briefly be in the courtroom with my family before we had to wait out the verdict separately.

It took about a minute for Judge Massei to declare the trial formally over. The time had come for the judges and jury to decide whom they believed. They exited single file through the door to chambers in the front of the courtroom. I stared at the door after it closed, wishing I knew what was going on behind it.

Then the prison van took me back to Capanne. I felt completely helpless, pointlessly thinking about what I should have said in my plea.

Back in my cell, I paced, sat on my bed, paced, sat. I tried to talk with my cellmates, Fanta and Tanya, but I was unable to concentrate on anything they were saying.

They were prepping me on all the superstitions I had to remember when I came back with the good verdict—break my toothbrush in half and throw it away outside the prison, with my hairbrush and the shoes I wore most often. This meant I wasn't coming back. "Just before you get in the car, remember to brush your right foot along the ground," Fanta said. "It means you're promising freedom to the next prisoner."

My head pounded as I shot from excitement to terror and back again—and again. My brain bounced between *Please, please, please* and *Finally, finally, finally—THE END*.

Besides my cellmates, Laura was the only person I could stand to see. She came during *socialità* and made chicken with mushrooms for dinner. I ate one bite.

I planned to give my pans, pots, and clothes to Fanta and Tanya.

I told Laura, "I want you to have my bedsheets."

"That will be great, Amanda," she said, "but don't promise me anything until we know what's going to happen."

"I'm going to write you, Laura," I told her.

"I hope so," she said. "But let's just wait and see."

After dinner Tanya turned on the TV. Every channel was talking about my case: The big day! The world is hanging on, waiting to see what the decision will be in the "Italian trial of the century." Raffaele and Amanda have been charged with six counts. Meredith's family will be there to hear the verdict. Amanda's family is waiting in the hotel. The Americans believe there's no case, but the prosecution insists that Meredith's DNA is on the murder weapon and Raffaele's DNA is on Meredith's bra clasp. The prosecution has condemned the American media for taking an incorrect view of the case.

The people on TV dramatized it: the lives of two individuals—will they walk free or spend the rest of their lives in prison? And on another channel: It's a question of whether Amanda goes free or gets *ergastalo*—"life imprisonment."

Tanya gasped. "What do they mean?" she asked.

Manuela Comodi, the co-prosecutor, had called for a life sentence, but it was as if I didn't understand how that related to me. I said, "Yeah, they asked for life."

"They're going to try to do that?" Far more than I, Tanya realized what was at stake. She was fidgeting.

In Italy, a life sentence means no parole. The next-lowest option, thirty years, offers the possibility of parole after twenty years.

"It's going to be okay!" I said. "Just calm down!"

A life sentence couldn't happen. *I have to be acquitted!*

The guards stopped by from time to time to see how I was doing.

I kept going back and forth from my bed to my locker to do an inventory of my things. Were the books, clothes, and papers I wanted to take out with me ready? Were all my letters organized in a folder?

Night fell, leaving the air outside damp and cold. Hours passed. I felt tingly, buzzing beneath my own skin. The verdict had to be coming soon.

Finally, I climbed into bed wearing everything but my shoes. I lay in the dark cell, which was illuminated only by the TV still talking about me and my future.

Chapter 29

December 4, 2009

I t was just after 11 P.M. I lay in my cot thinking, *Maybe it won't even happen tonight,* when a guard came by. "Amanda, are you ready?" she called, putting her key in the lock.

I jumped out of bed and started to smooth my sheets. "No!" Tanya and Fanta shouted. "Don't do that! You have to leave your bed unmade. It's good luck! It means you're not coming back."

I put on my shoes, took a quick look around, and walked out, leaving my cellmates standing at the *cancello*—the cell's gatelike door—watching me walk down the hall.

It was surreal to go outside at this hour. Since my arrest, the only time I'd been out later than 3 P.M.—the end of *passeggio*—was on court days. Even then I was usually back in my cell before dark. I'd only felt the night air and seen the moon through my window.

It was damp and frigid, the full moon obscured by fog.

This is the last time, I thought as I climbed into the van, waiting for the guards to slam first the bars and then the double doors in back. After dozens of these trips, I no longer paid attention to the routine. But tonight I felt I had to take it in. *This is it! Never again!* I'd be coming back to Capanne to gather my things in a squad car.

My heart was thudding, and the only thought looping through my mind was the same one I'd been saying to the universe all day. *Please, please, please, please.* I was shaky with nerves and cold. But underneath the anxiety was a hard kernel of certainty. It was almost as if I were in on a secret that no one else knew. *I'm getting out! I'm going home!*

Usually the drive into town made me nauseated, but this time I didn't focus on the van's swaying. I had a physical memory of every curve in the road. I got frustrated when the guard closed the shade between the prisoners' compartment and the front seat, so I couldn't see out. I always strained to see farmers working green fields or the stretch of road where sunflowers grew—a world saturated with color and filled with hope instead of the beige-and-gray universe I inhabited at Capanne.

But tonight it didn't matter. I was lost in my thoughts. *The jury must have gone over all the evidence and seen that it doesn't fit. Raffaele and I couldn't have killed Meredith. The judge would read the counts and announce* "assolta"—"acquitted." *Anything but* "colpevole"—"guilty."

Some days it had seemed I waited in the van forever before being taken inside the courthouse, but everything was happening quickly now. I was whisked inside and up the stairs. My sisters Ashley and Delaney were standing by the double doors as the guards propelled me past. They each called out, "I love you, Amanda!" in heartbreakingly sweet voices.

I could have touched them if the guards had let me. I was that close.

The Hall of Frescoes had been transformed. All the chairs had been taken out, and hundreds of people were standing jammed together. The room was as quiet as it was packed. No journalists called out to me. Everyone was silent. Expectant.

I'd just seen Mom, Dad, Chris, Cassandra, my aunt Christina, and Deanna that morning, and here they were again, standing in a line, smiling, everyone mouthing the same words: "I love you, I love you."

I took my place between Carlo and Luciano, squeezing Luciano's bearlike hand. "*Coraggio,*" he whispered, squeezing back.

It was four minutes past midnight. The court bell rang once. The secretary announced, "*La corte,*" for the last time. As the judges and jurors filed in, it was as though everyone in the courtroom strained forward, all the energy and nervousness and anticipation driving to the same point in time and space.

Each of the six counts against me—murder, carrying a weapon, rape, theft, simulating a burglary, and slander—had been assigned letters A through F, in that order.

In the seconds before the judge started reading, I felt both a downward tug in my stomach and wooziness in my head that made me feel as if my body were being pulled apart. It was all going to be over. *Please, please, please.*

"On the counts of A, B, C, E, and partially for count D"— Judge Massei began reading Raffaele's and my verdicts simultaneously, his voice flat and so quiet that I struggled to hear, willing him to say "*assolta*"—"the defendants Sollecito, Raffaele, and Knox, Amanda, are found . . ."

"No!" someone behind me wailed.

"*Colpevole,*" Judge Massei said. "Defendant Knox, Amanda, is also found *colpevole* for count F."

Flattened by the words, I could no longer stand. I fell against

Luciano, burying my head against his chest, moaning, "No, no, no!"

I didn't hear the judge say, "I'm granting the *attenuanti*"— "extenuating circumstances," meaning a lower sentence. "I'm sentencing Knox, Amanda, to twenty-six years and Sollecito, Raffaele, to twenty-five years. This court is adjourned."

My life cleaved in two. Before the verdict, I'd been a wrongly accused college student about to walk free. I was about to start my life over after two years.

Now everything I'd thought I'd been promised had been ripped away.

I was a convicted murderer.

I was less than nothing.

I didn't hear people cheering or jeering. Some were calling me an assassin. Others were calling for my freedom. The only sounds I picked up, above the chaos, were my mom's and Deanna's sobs rising up behind me and smothering me in pain.

Then the guards on either side of me lifted me under my arms and carried me out of the room. Ashley and Delaney must have been standing in the same spot I'd seen them before, waiting to hug and kiss me in celebration, but I could not see through my tears.

Carlo stopped us just before we started down the stairs. He was breathless. "I'm so sorry! We're going to win! We're going to win. Amanda, we're going to save you. Be strong."

It was only a second. And then we were gone. Instead of putting me in the tiny holding cell where I usually ate lunch or waited for the van, the guards sat me in a chair. I was moaning, "No, no, no," hysterically. Raffaele was beside me, saying, "Amanda, it's okay, it's okay."

One of the guards kept saying, "Come on. Be a good girl. Hold on. It's going to be okay."

I kept crying, "It's impossible, it's not fair, it wasn't true, I need to go home."

They led me outside to the van and slammed the barred door.

As we were pulling out, I took one look outside. The guard driving the van hadn't pulled down the shade, and I could see the cameras flashing. Then I slumped over in my seat and wailed, gasping for breath.

My sentence was all over the news.

"Twenty-six years is a strange sentence," one of the guards said. "If they wanted to get you they would have said, 'Life.' It's almost like they were trying to give you hope for the future. You have such good behavior. In ten years you'll be able to work outside the prison during the day."

They were trying to reassure me.

But I could not be comforted.

Part Three

CAPANNE II

Chapter 30

December 2009–October 2010

A
t Capanne there were two kinds of suicide watch.

The first was for people who had previously tried to kill themselves or were mentally ill. They were put in a bare cell with a guard stationed in front of the door 24/7. The second kind was for prisoners who had no history of suicide attempts or mental health issues, just a good reason to try to kill themselves. I was in that group.

It meant a guard would look in on me every five minutes.

I wasn't suicidal, but my insides had already died.

My first stop on arriving back at Capanne from the courthouse was an office just inside the door to the women's prison. "You're going to be okay," said Lupa, the guard who came into my cell on my first day in prison and hugged me. "Your lawyers will appeal the decision, and something good will come of it. You'll see."

You're wrong, I thought, *I've been convicted of murder. I will never again be okay. Nothing good will come from this.*

The realization of what had happened at the courthouse took the air from my lungs and the heat from my body. I shivered so much an *agente* brought me hot milk, like you'd give a baby. All I could think was, *What the fuck do I care about warm milk? I don't want milk, I want to go home.*

Besides Lupa, a couple of other guards stayed with me. One was from the women's ward, and the other had escorted me on some of my trips to the courthouse. Whether or not they thought I was innocent, they were trying to be nice. But I didn't want to be consoled by them. I said, *"Grazie,"* over and over. But I was sickened, hollowed out, crushed.

I remember turning six. Waiting to turn sixteen. I thought it would be cool to turn twenty-one in Italy. I'd never thought about twenty-six years. Twenty-six was an old number. It wasn't a measurement I ever used. My mom and Chris had been married eight years. My dad and Cassandra, twenty years. I remembered seeing a bottle of whiskey that said "Aged Twenty-Five Years." But I'd never thought about twenty-six years. It was a year older than the whiskey and four years older than me. I'd be forty-eight when I got out of jail, one year older than my mom's age on the day I was sentenced. I divided twenty-six by what I knew. Twenty-six years was thirteen times as long as I'd been in prison.

I couldn't stop doing the math. Each permutation added up to the same thing: I had been convicted of murder.

"Please, don't take me to my cell," I pleaded.

I didn't want to have to tell Tanya and Fanta or to cry in front of them. Two *agenti* sat with me for about an hour. Then their shift was over. They had to go home, and so did I. Only I couldn't.

I wept until I felt I was suffocating. Thinking about my family—how much they had sacrificed and how disappointed they were—was the worst. I wanted to be with them, to hold them and be held. Instead, I sat in front of the *ispettore*'s desk lost in anguish.

"Is there anything we can do for you, Kuh-nox?" she asked. It was the same orange-haired *ispettore* who'd called me out on the local newspaper article about Cera that caused me to be treated as an *infame* by the other prisoners.

I knew I was too fragile to withstand any more hostility. I was lucky our cell had dwindled to three women, but with two empty beds, anyone could move in and anything could happen. The constant turnover of cellmates made my tiny life even smaller.

"Can you possibly put me on the list for a two-person cell instead of the five-person cell?" I asked, sniffling. "That would mean a lot to me." It was all I had. Begging for a better cell. It had come to this. This was my new life.

I was in a position to ask. Twenty-six-year sentences were uncommon in Italy, especially at Capanne, which usually housed petty criminals and drug dealers serving sentences of a few months to a few years. After twenty-five months, not only had I earned seniority—I'd been there longer than almost everyone else—but I had a reputation as a model prisoner.

Back in my cell, Fanta and Tanya hovered near me, unsure of what to say—or whether to say anything. Tanya finally hugged me, and Fanta said, "I'm so sorry. We watched the reports on TV."

About an hour later, a guard came to the *cancello*. "Get your things, Kuh-nox. You're moving in with Laura."

This was an unheard-of kindness. It meant they'd moved my friend Laura's cellmate out just to accommodate me.

After so much had gone against me, Laura was the right cellmate at the right time.

When we first met, we'd entertained each other making light of prison's darkest aspects—being subjected to daily strip searches by *agenti*—and joked endlessly about the ordinary things we missed most from real life. My answer? Sushi. We sang "The Star-Spangled Banner" together each morning in a show of patriotism and homesickness.

Laura grew up in Ecuador in an American military family. She was years older than my mother, and I was younger than her two daughters, but we'd become friends. Besides Don Saulo, she was the only person at Capanne whom I trusted.

In Quito, where she lived, Laura had dated an Italian who invited her to Naples for vacation and bought her a new suitcase. When she landed at the Aeroporto Internazionale di Napoli, it was not her boyfriend who met her plane but the customs police. They arrested her for the cocaine they found sewn into the luggage's lining. The boyfriend, it turned out, had not only turned her into a drug mule, but had lied about his name. He was untraceable. She was sentenced to nearly five years in prison.

Laura held herself as erect as a flagpole and seemed aristocratic even in her sweatpants. She respected the guards but didn't lower herself. She had a job in the kitchen, but as the economy, and the prison pay, shrank, she politely said, "My service is worth more, and I won't work for this little."

They promoted her from scullery maid to chef, and gave her a raise.

Laura was the perfect person to teach me how to stand up for myself.

"You come first, second, and third, then everyone else," she told me when I agonized over whether I was writing my family and friends enough or whether I was treating other prisoners with the right balance of generosity and restraint.

She reminded me, "Amanda, it's okay to say no. Prisoners are asking you to do them favors they know jeopardize you."

The prison was divided into different sections, and we were forbidden to pass things—coffee, cigarettes, anything—from one section to another. If I got caught, it would cost me months I could otherwise take off my sentence for good behavior. "They can go without coffee," Laura said. "You're thinking too much about other people at your own expense, and you're stretching yourself too thin."

When I'd beat myself up over past mistakes, she reminded me, "You're a much better person than you give yourself credit for."

And when I obsessed over whether certain prisoners might pick a fight with me, she scoffed and said, "If they bark, they won't bite."

Laura's approach was brilliant—if you aren't afraid, it's less likely someone will attack you. "They can smell your fear," she said. "You see how no one messes with me? It's because I'm not afraid of them."

I was never as bold as Laura, but her tough-love encouragement helped me gain self-confidence. She helped me realize that despite my mistakes and unmet expectations, I was a good person.

And while I learned a lot from her, this wasn't a one-way relationship. I made her laugh, a real coup, because she didn't suffer fools. She said I spoke my own goofy language; she called it "Amandish." When I started acting too chipper and silly—impersonating people or making up stories like you'd tell a child at bedtime—she'd tap an imaginary "weirdness gauge" on the wall above her head and say, "You're off the scale, Amanda!"

Laura's friendship and a few others held me together.

Rocco Girlanda and Corrado Daclon came to visit me on the Sunday after my conviction. As the president and vice president, respectively, of the Italy-USA Foundation, they said they wanted to help me. Rocco was a middle-aged conservative politician with a boyish face and smile. Corrado was a talkative professor of economics. Initially, I was suspicious of them. I received, and threw away, plenty of mail from the morbidly curious. Later, the two men told me that they had arrived with trepidations of their own. They were relieved to find that I wasn't as I'd been billed.

Their friendship uplifted me. They visited at least once a month and sent me books once a week. They gave me a Mac computer and bought me an iPod as a birthday present—and somehow they managed to convince the director of the prison to let me use both.

———————

Besides Laura and Don Saulo, there was one other prison friendship I prized. It was with a toddler named Mina, whose mother, Gregora, was at Capanne for stealing.

Completely uneducated, Gregora couldn't name the year, date, or time. She couldn't read, write, add, or subtract. She didn't know how old Mina was, only that she'd been born when it was cold outside.

Fanta had introduced us. "Gregora needs someone to write letters for her," she said. I laid down the same rules with her as I did with everyone I helped write letters: "I won't think up the words, but I'll take down what you want to say. You talk, I write."

Since Gregora and Mina were in the nursery ward, I saw them during *passeggio*. A high wall of bars was all that divided our outside areas. Gregora would slip me pen and paper, and I'd dedicate the first half hour of our afternoon outdoor time to her. I'd pause to talk to Mina, who always played by herself. She toddled around in what seemed to be self-imposed silence, gesturing to communicate. Having spent most of her life in prison, Mina never stepped through a doorway without permission, turning her hand to signify a key in a lock. Serious and suspicious of strangers, she seemed to have an ancient soul—weary, alert, and wise. Sometimes she'd bring over a timeworn doll and cradle it for me, nodding her head and meeting my eyes, as though I could pour out my heart to her and she'd understand.

Mothers and their children were also allowed to attend Don Saulo's group activity time. Mina sat on my lap during movies, let me carry her around the room, and chose me as her dance partner when Don Saulo played religious music. She liked to switch shoes with me. She'd hang her own tiny, red plastic ones on my toes and clomp around in mine.

One afternoon Gregora ran up to the bars outside, calling, "Amanda! Amanda!"

I came over, expecting Gregora to hand me the latest letter from her husband, a prisoner on the men's side. Instead she whispered, "Listen!"

I looked around to see Mina playing by herself in the middle of the yard.

"It's the song you sing in church!" Gregora cried.

"Ave-sha-om-ahem . . ."

I could hear a tinny, high-pitched voice squeaking out a melody.

"Hevenu shalom alechem"—"May peace be with you." It was one of the prisoners' favorite songs during Mass, which I accompanied on guitar.

It can't be Mina! I'd always imagined that if she ever talked, or sang, her voice would be husky and deep, like an old woman's.

That's how she carried herself. Hearing her peep out a song in a tiny baby voice clutched at my heart.

———————

I kept my promise to myself. After my conviction, I got the first appointment I could with the volunteer hairdresser. "Cut it off," I said.

The woman next to me, her hair wrapped in tin foil, gasped, "You're crazy." The hairdresser met my eyes worriedly in the mirror.

"Are you sure?" she asked.

"Yes, just do it," I said more forcefully than I'd meant to.

I ended up with a crude, boyish cap of a cut. I'd fallen into magical thinking, believing that short hair would transform me, that this protest in a teacup would somehow make me feel better about my conviction or turn me into someone else.

What it did was earn me a trip to the psychiatrist's office. "You know, people make drastic alterations only when they're asking for attention," she chided.

"That's not true for me," I responded, irritated. "I just want to be left alone. What I do with my hair is my business."

"Have you thought any more about taking an antidepressant?" she asked.

"No, thank you," I said curtly.

"When you eventually get out of here, you're going to need a lot of help—psychologists at the very least," she said.

I hated people lecturing me as if they had a clue about how I felt now and would feel in the future. *No one can understand what I think. Even the people who love me best can't completely identify with me.*

But talking to them was holding me together. I lived for prison visits and my once-a-week Saturday-night-at-seven phone call to friends and family in Seattle. For those who couldn't visit me in Perugia, it was my only connection to their voices. On Saturdays I'd count down in my head, and at exactly ten minutes before seven o'clock, I'd shout, "*Agente*, phone call!" One night I yelled and no one came. My call would be connected at seven on the dot. "*Agente!* Phone call!" No answer. "*Agente*, phone call!" No answer.

I crumpled onto the floor and rolled into a ball, weeping and screaming. I felt like a dog in a kennel, behind bars, howling for help. I was crying out for someone, and no one came. My family was waiting on the line, and the phone was only a few paces away. It was as close as I have ever come to a breakdown.

I was still screaming when the guard came at 7:30.

"I was downstairs," the *agente* said. "Sorry."

I didn't know then that the prison budget had been cut and that guards who used to cover one floor now had to cover two. I was so angry I doubt I would have cared. I didn't get my hopes up for anything in prison. I didn't expect anyone to do anything nice for me. But I counted on my phone call. I stayed out of trouble. I helped wherever I could. And now fourteen days would have passed before I could talk to my family in Seattle again. I couldn't count on justice, and I couldn't count on people.

Looking back, I thought how stupid I was in November 2007 when I'd first been arrested. I thought I was a special case and would be kept in prison for only a few hours—a few days at most—for my "protection." When the investigation started, I thought it was just a matter of time until the prosecution realized I'd been wrongly accused. When I was being tried, I was sure I wouldn't be convicted. But I had reached the end of the line. This was now my life. I was not special. In the eyes of the law, I was a murderer.

As Lupa said, my lawyers would obviously appeal my conviction. But I couldn't count on the Court of Appeals to free me. My case, tried daily in the media, was too big and too notorious. It was awful to hear that strangers believed I had killed my friend. That feeling was compounded when, about three weeks after Raffaele and I were convicted, the appeals court cut Rudy Guede's sentence nearly in half, from thirty years to sixteen. Meredith's murderer was now serving less time than I was—by *ten years! How can they do this?!* I raged to myself. *It doesn't make sense!* The unfairness of it burned in my throat.

Guede's fast-track conviction for murder and rape in collaboration with others had earned him the maximum. The appeals court had also found him guilty on the same count. But the prosecution's new view—and the reason for the reduced sentence—was that Guede had not had the knife in his hand, and therefore had played only a supporting role, more responsible for Meredith's rape than for her murder.

Two weeks into the new year, I was called to the first floor to sign a document. I assumed it would confirm my conviction. I thought, *I'm already living this god-awful reality every day. I don't need a piece of paper to make it official.* But when the emotionless guard pushed the paper across the desk, I saw, to my astonishment, and outrage, that it was a new indictment—for slander. For telling the truth about what had happened to me during my interrogation on November 5–6, 2007.

In June 2009, I testified that Rita Ficarra had hit me on the head to make me name Patrick.

I also testified that the police interpreter hadn't translated my claims of innocence and that she'd suggested that I didn't re-member assisting Patrick Lumumba when he sexually assaulted Meredith.

According to Prosecutor Mignini, truth was slander.

All told, the prosecution claimed that I'd slandered twelve police officers—everyone who was in the interrogation room with me that night—when I said they'd forced me to agree that Meredith had been raped and pushed me into saying Patrick's name.

It was my word against theirs, because that day the police ap-parently hadn't seen fit to flip the switch of the recording device that had been secretly bugging me every day in the same office of the *questura* leading up to the interrogation.

Making myself read to the end, I saw that the lawyer repre-senting the police department was Francesco Maresca. He was also the Kercher family's civil lawyer.

Mignini and his co-prosecutor, Manuela Comodi, had signed the document. The judge's signature was also familiar: Claudia Matteini, the same woman who'd rejected me for house arrest two years earlier because she said I'd flee Italy.

I hadn't expected this maneuver by the police and prosecu-tion, but it now made sense. They couldn't admit that one of their own had hit me or that the interpreter hadn't done her job. Above all, they couldn't admit that they'd manipulated me into a false admission of guilt. They had their reputations to uphold and their jobs to keep.

I'd calculated that I could be released in twenty-one years for good behavior. Now this looked unlikely. If I were called to tes-tify in the slander trial, I'd have to restate the truth: I had been pressured and hit. They'd say I was lying. If the judges and jury believed the police, that would wipe out my good behavior and add three years to my jail time.

Could Mignini, Comodi, and the whole *questura* keep going after me again and again? Would I be persecuted forever?

The indictment was a dark reminder of how completely vulnerable I was. Not only had the prosecution successfully had me convicted for something I hadn't done, but also legally, my word meant nothing. I was trapped.

And so angry. I'd never felt so consumed by raw, negative emotion as I did then. I had to turn my thoughts away from it. For the first time, I was afraid of the spiteful, miserable bitterness I felt.

Incredibly, a month later, Prosecutor Mignini was convicted for abuse of office and sentenced to a sixteen-month suspended sentence for his part in the Monster of Florence case. He was accused of having used his authority to intimidate and manipulate people. By the time I was convicted, there was no question that he'd also manipulated me. That case is currently on appeal.

My sense of doom was growing. With the prosecutor's verdict coming so close to my own, it seemed that they'd waited to convict Mignini until he'd convicted me. Sitting in my barely heated prison cell on that frozen January day, I believed that the Italians had made a mockery of the word *justice*.

———

As the months went by, I realized that I hadn't just been convicted of murder—I'd been sentenced to a life apart from the people I loved.

In a practical sense, my innocence didn't matter anymore. Whether or not I belonged there, prison was suddenly my entire world.

I'd always tried to fill my days there with mental and physical exercise—*in the meantime*, I told myself, *until I go home*. I'd taught myself Italian and I was healthy.

After my conviction, my sense of purpose became my life raft. I clung to purposefulness. It was the only thing that allowed me to maintain my relationships, my humanity, my sanity. I was obsessed with making each day count. The one thing I couldn't tolerate was wasting my life in jail.

Prison officials started calling me to be an interpreter for anyone who didn't speak Italian—even if that other language was Chinese and I had to point to words in the English-Chinese dictionary I happened to have.

As I did for Mina's mom, Gregora, I helped prisoners write letters, legal documents, grocery lists, and explain an ailment to the doctor. The Nigerian women treated me as an honored guest, setting me up at a table and offering tea and cake as they dictated to me. This was my way of being part of the prison community on my own terms, of trying to find a good balance between helping others and protecting myself. No matter how much I was hurting, I didn't think it was right to ignore the fact that I could help other inmates with my ability to read and write in both Italian and English.

At bedtime each night, I made a schedule for the next day, organized task by task, hour by hour. If I didn't cross off each item, I felt I'd let myself down. I wrote as much as I could—journals, stories, poems. I could spend hours crafting a single letter to my family.

I thought about what I wanted to say to whoever came to visit me that week, and the message I wanted to convey on my weekly phone call home.

I became a purposeful reader. I already preferred Franz Kafka to Jackie Collins, but now I was drawn to books with characters who were isolated, lost, or grieving in a surreal, existential way— Fyodor Dostoyevsky's *The Possessed*, Aleksandr Solzhenitsyn's *The First Circle*, Marilynne Robinson's *Housekeeping*, and Umberto Eco's *The Island of the Day Before*.

I read a lot! And I often felt more solidarity with the characters and the writers who created them than I did with the real people I knew.

Since my arrest, the prison psychiatrist had often asked me if I harbored suicidal thoughts. Her question always struck me as strange—as did the suicide watch I was put on the night of my verdict. How could people kill themselves? *There's always hope. There's always something to be gained in life.* No matter what, my life meant something to me and to my friends and family. *I could never consider such a thing.*

But I struggled with a way to come to terms with the fact that my life was encapsulated within these gray walls. And I started to understand how you could feel so locked inside your own life that you could be desperate to escape, even if it meant that you'd no longer exist.

The ways other prisoners had tried to kill themselves were well known—and I imagined myself trying them all.

There was poisoning, usually with bleach. Swallowing enough and holding it in long enough was painfully difficult. Usually the vomiting would attract the attention of the guards too soon, and then they'd pump your stomach. It seemed an agonizing way to go if success wasn't guaranteed.

There was swallowing shards of glass from a compact mirror or a broken plastic pen, hitting your head against the wall until you beat yourself to death, and hanging yourself.

But the most common and fail-safe method of suicide in

prison was suffocation by a garbage bag—two prisoners on the men's side did this successfully while I was there. You could even buy the bags off the grocery list. You'd pull the bag over your head, stick an open gas canister meant for the camping stove inside, and tie the bag off around your neck. The gas would make you pass out almost instantaneously, and if someone didn't untie the bag immediately, that was it.

Less effective but, I thought, more dignified was bleeding yourself to death. I imagined it would be possible to get away with enough time in the shower. The running water would deter cellmates from invading your privacy, and the steam would fog up the guard's viewing window. I imagined cutting both my wrists and sinking into oblivion in a calm, quiet, hot mist.

I wondered which straw would need to break for me actually to do any of these. What would my family and friends think? How would the guards find my body?

I imagined myself as a corpse. It made me feel sick, not relieved, but it was a fantasy I had many times—terrible, desperate recurring thoughts that I never shared with a soul.

I also imagined what it would be like to live a life not inside prison. *If all this hadn't happened, where would I be?* I pictured myself being a regular person—going to the grocery store, getting coffee at Starbucks, having lunch with my mom, rock climbing. I'd get lost in memories of when I was younger—walks with Oma and July Fourth fireworks with Dad; playing football with my friends or helping Madison develop photos; bicycling with DJ and taking long walks with my friend James.

I thought about how much I wanted to get married and have kids. *If I get released on good behavior when I'm forty-three, I can still adopt.*

Other prisoners would say, "You're lucky you're in prison

while you're young, because you're going to get out and stand on your own two feet."

I thought, *What the fuck are you talking about? How will I know how to live my life? I won't have been given a chance to be an adult.*

I started having conversations with myself as if I were talking to a younger sister. I told her, *Don't rush, keep your eyes open, observe things, don't be so insecure. You're just fine.*

She'd tell me, *Stop being so hard on yourself.*

It calmed me, but I started worrying that I was going crazy. *Is this one of the steps people take on the way toward losing their mind?*

Optimism had been my way of life, and it still was for my mother, who continued to insist that I'd be freed. But optimism had not saved me. I could picture myself growing old in prison, losing everything I'd ever hoped for in life, and one day returning to the world a ghost of a person, without anyone capable of understanding me. I thought I'd look like a smaller version of Laura with brown hair—maybe because she was close to the age I would be by then.

My feelings of loss only made it worse for me the day Mina was taken away from Gregora and put in an orphanage. Although Gregora didn't know when Mina was born, prison officials had decided that she must be three. From then on, a social worker would bring Mina to visit her mother for an hour a month. I couldn't decide whether I identified more with the child I'd come to love or with her inconsolable mother. But I was sure of one thing: prison tore families apart, and they could never be stitched back together.

My mom couldn't accept my sadness. She wrote, and talked to me, many times about how scared she was for me. "You're changing, Amanda," she said. "You're not sunny anymore. I hope when you get out you can go back to being the happy person you were."

"Mom," I wrote back, "good things don't always work out for good people. Sometimes shit happens for no reason, and there's nothing you can do about it."

I had to consider the worst- and best-case scenarios to be able to maintain my emotional stability. I was no longer a tourist, waiting to be allowed to go home. I was a prisoner. I was trying to figure out how to be happy even if I weren't freed. I was preparing for life in prison. The thing that scared my mom was that I wasn't completely focused on getting out, which she saw as giving up.

I wasn't. I hadn't lost hope, but I wasn't banking on it, either.

Mom couldn't understand what I was trying to say: that it was important not to be optimistic all the time but to come to terms with reality, the good and the bad, and to create something positive from it. She took this to mean that I'd become a pessimist. It was hard for us to communicate with each other about this.

My conviction and the ways I tried to deal with it created a divergent path that made me afraid I could lose my family. Not literally, but in my soul. Was I becoming someone different, someone they couldn't reach?

I never thought they'd abandon me. Remarkably for a stepfather, Chris had kept his word and essentially moved to Perugia, living there for months at a time, and there was always someone to visit me during the twice-weekly visitation schedule. At home, Oma had started lighting a candle to represent me at each family gathering, and Deanna told me in a good-humored way, "Look how crazy we all are! We light a candle and pretend it's you!"

But I was consumed by the question of how long they could,

or should, keep up the back-and-forth between Seattle and Perugia. *What is my family going to do? What about work? What about life?*

It was the same for my friends who were graduating that spring. *How do I tell them that I live here in prison now, and that they need to move on with their lives—without me? Can I tell them just to forget about me for 26 years?*

I desperately didn't want to be forgotten. But more than worrying about the logistics of such a life, I was terrified that we were coming to a point where we wouldn't understand one another. They still had the right to choose what to do with their lives; they had freedom. I didn't. I was at the mercy of my wardens. I worried that my new prison identity wouldn't make sense to them, and my mom was evidence of that. If enough time passed, we'd be speaking two different languages—and it would have nothing and everything to do with their English and my Italian.

Chapter 31

———————

November–December 2010

S itting beside me in the visitors' room at Capanne, my
friend Madison reached over and brushed my cheek. I
flinched.

"Baby, don't worry. It's just an eyelash," she said.

My skittishness horrified me. "I guess I'm just not used to
people touching me anymore."

In one of the few happy surprises of my then-three years in
prison, Madison had moved from Seattle to Perugia to be near
me—arriving in November 2010, a few days after Laura handed
me her bedsheets, hugged me tearfully, and left for a halfway
house in Naples.

One friend couldn't replace the other. Laura had nurtured me
on the inside. Seeing Madison twice a week for an hour gave
me heart—and, often, valuable information. Distraught after a
bad breakup, Madison embraced the Mormon philosophy she'd
grown up with: "forget yourself and go to work." Keeping me
grounded and hopeful was her work. But she also needed a paying job. A photographer who didn't speak Italian, she turned to
Rocco and Corrado. With their recommendation and her port-

folio, she landed a gig taking pictures for a local newspaper. In her spare time, she and an interpreter friend talked to journalists, lawyers, and people on the street about my case. Maybe one of those interviews would give Carlo, Luciano, and me an extra fingerhold to help pull me out of this hell.

After I was convicted, my family, my lawyers, my friends, other prisoners—even, bizarrely, prison officials—tried to console me by telling me that I'd surely have my sentence reduced, if not overturned, on appeal. Rocco and Corrado assured me that in Italy about half the cases win on appeal.

The old Amanda would have appreciated the possibility.

But I'd been burned so often I was terrified. Why would the Court of Appeals make a different decision from the previous court? Or from the pretrial judge? Both had accepted the prosecution's version. With my case, the Italian judicial system was also on trial. My story was well known, and the world was watching. It'd be difficult for the judicial authorities to back down now.

One thing had changed: me. *I* was different. In the year since my conviction I'd decided that being a victim wouldn't help me. In prison there were a lot of women who blamed others for their bad circumstances. They lived lethargic, angry lives. I refused to be that person. I pulled myself out of the dark place into which I'd tumbled. I promised myself I'd live in a way that I could respect. I would love myself. And I would live as fully as I could in confinement.

The questions and choices I made during the first trial ate at me. *What if I'd spoken up more, clarified more when other witnesses took the stand, pleaded my innocence more forcefully? Would it have made a difference?* I'd waited for the jury and the world to realize that there was no evidence against me. I wasn't going to make the same mistake twice.

Though I trusted my lawyers completely, this time I wanted to be involved in every decision. I owed it to myself. I couldn't survive another guilty verdict if my team and I overlooked a single speck of favorable evidence.

Once I started thinking about what might be possible, nothing seemed out of reach. Should I write to the new judge? The U.S. secretary of state? Why not the president?

Rather than write, I read. The 407-page report from Judge Massei explained why we'd been convicted and how Raffaele, Guede, and I had murdered Meredith.

The supposed motive was as far-fetched as a soap opera plot. "Amanda and Raffaele suddenly found themselves without any commitments; they met Rudy Guede by chance and found themselves together with him at the house on the Via della Pergola where . . . Meredith was alone," Massei wrote.

The judges and jury hypothesized that Raffaele and I were fooling around, and that Guede started raping Meredith because we turned him on. Instead of helping Meredith, we inexplicably and spontaneously joined Guede, because it was "an exciting stimulant that, although unexpected, had to be tried," he wrote. "[T]he criminal acts were carried out on the force of pure chance. A motive, therefore, of an erotic, sexually violent nature which, arising from the choice of evil made by Rudy, found active collaboration from Amanda Knox and Raffaele Sollecito."

The report rejected the prosecution's claim that Meredith and I had had a contentious relationship. The judge wrote "the crime that was carried out . . . without any animosity or feelings of rancor against the victim . . ."

They allowed that there was no evidence of contact between Guede and me—no e-mails, phone calls, or eyewitnesses. They discounted the testimony of Hekuran Kokomani, the witness from the pretrial and the trial who said he threw olives at me and who "identified" me by the nonexistent gap between my teeth. And they conceded that Raffaele and I were not likely killers. Rather we were "two young people, strongly interested in each other, with intellectual and cultural curiosity, he on the eve of his graduation and she full of interests . . ."

Nonetheless, the report claimed, Raffaele and I "resolved to participate in an action aimed at forcing the will of Meredith, with whom they had, especially Amanda, a relationship of regular meetings and cordiality, to the point of causing her death . . . the choice of extreme evil was put into practice. It can be hypothesized that this choice of evil began with the consumption of drugs which had happened also that evening, as Amanda testified."

It continued: "Therefore it may be deduced that, accustomed to the consumption of drugs and the effects of the latter, Amanda Knox and Raffaele Sollecito participated actively in Rudy's criminal acts aimed at overcoming Meredith's resistance, subjugating her will and thus allowing Rudy to act out his lustful impulses . . ."

Another factor, the judge wrote, was that Raffaele and I read comic books and watched movies "in which sexuality is accompanied by violence and by situations of fear . . ."

He brought up the disputed theory that Raffaele's kitchen knife was the murder weapon, in addition to a new theory that I'd carried the knife in my "very capacious bag." Why would I? "It's probable, considering Raffaele's interest in knives, that Amanda was advised and convinced by her boyfriend, Raffaele Sollecito, to carry a knife with her . . . during the night along streets that could have seemed not very safe to pass through at night by a girl."

The lining of my bag wasn't cut. The police found no blood in my bag. How can I prove what I didn't do?

The prosecution had based their case on misinterpreted and tainted forensic evidence and had relied heavily on speculation. But Judge Massei's faith was blind. Patrizia Stefanoni would not "offer false interpretations and readings," he wrote.

The appeal wouldn't be a redo of the first trial. Italy, like the United States, has three levels of justice—the lower court, the Court of Appeals, and the highest court, the Corte Suprema di Cassazione, their version of our Supreme Court. The difference is that, in Italy, someone like me is required to go through all three levels, all the way to the Cassazione, whose verdict is final.

Cases often take turns and twists that would surprise and unsettle most Americans. Even if you're acquitted at level one, the prosecution can ask the Court of Appeals to overturn the verdict. If the appeals court finds you guilty, it can raise your sentence. Or it can decide that a second look is unnecessary and send you on to the Cassazione for the final stamp on the lower court's decision—in Raffaele's and my cases, to serve out our twenty-five- and twenty-six-year sentences.

At each level, the verdict is official, and the sentence goes into immediate effect unless the next court overturns it.

In Italy's lower and intermediate levels, judges and jurors decide the verdict. And instead of focusing on legal errors, as we do in the United States, the Italian appellate court will reopen the case, look at new evidence, and hear additional testimony—if they think it's deserved.

In our appeal request, we asked the court to appoint independent experts to review the DNA on the knife and the bra clasp, and to analyze a sperm stain on the pillow found underneath Meredith's body that the prosecution had maintained was irrelevant. In their appeal request, the prosecution complained about what they thought was a lenient sentence and demanded life in prison for Raffaele and me.

I read and reread the Massei report, looking for discrepancies and flawed reasoning. I'm not a lawyer, but I had an insider's perspective on the case, three years in prison, and eleven months in court. In one of Guede's depositions, he claimed I'd come home the night of the murder, rung the doorbell, and that Meredith had let me in. Obviously he didn't know it was our household habit to knock, not buzz. It was a little catch, but it was something my former Via della Pergola housemates, Laura and Filomena, could confirm.

Before arriving in Italy, Madison sent me lists breaking down the case by category and pressing me to consider it from different perspectives. Besides being a remarkable friend, Madison has, as part of her makeup, a stubborn, idealistic personality that insists on protesting wrongs and standing up for people she thinks need her help. I was lucky that she stood up for me.

For example, she wrote, "Witnesses: the prosecution knowingly used unreliable witnesses.

"Interrogation: the police were under enormous pressure to solve the murder quickly.

"There's a pattern of the police/prosecution ignoring indications of your innocence. This must be pointed out. You were called guilty a month before forensic results, you were still considered guilty even though what you said in your interrogation wasn't true, obviously false witnesses were used against you. The jury needs to know that you are being railroaded. How can you emphasize that? You can't just say 'I'm a scapegoat.' You must present a series of convincing points."

I knew that the most critical point was to be able to say why I'd named Patrick during my interrogation.

The prosecution and civil parties argued that I was a manipulative, lying criminal mastermind. My word meant nothing. The court would always presume I was a liar. If, in their mind, I was a liar, it was an easy leap to murderer.

I had been done in by my own words. I'd told the judges and jury things like "I didn't mean to do harm" and "You don't know what it's like to be manipulated, to think that you were wrong, to have so much doubt and pressure on you that you try to come up with answers other than those in your memory."

Thankfully Madison had researched the science on false confessions. She found Saul Kassin, a psychologist at John Jay College of Criminal Justice in New York. A specialist in wrongful convictions, he took the mystery out of what had happened to me.

Before my interrogation, I believed, like many people, that if someone were falsely accused, they wouldn't, couldn't, be swayed from the truth while under interrogation. I never would have believed that I could be pressured into confessing to something I hadn't done. For three years I berated myself for not having been stronger. I'm an honest person. During that interrogation, I had nothing to hide, and a stake in the truth—I desperately wanted the police to solve Meredith's murder. But now I know that innocent people often confess. The records kept of people convicted of a crime and later exonerated by DNA evidence show that the DNA of 25 percent of them didn't match the DNA left at the scene. The DNA testing showed that one in four innocent people ended up confessing as I did. And experts believe that even more innocent people confess, both in cases with and without DNA evidence.

According to Kassin, there are different types of false confessions. The most common is "compliant," which usually happens when the suspect is threatened with punishment or isolation. The encounter becomes so stressful, so unbearable, that suspects who know they're innocent eventually give in just to make the uncomfortably harsh questioning stop. "You'll get thirty years in prison if you don't tell us," says one interrogator. "I want to help you, but I can't unless you help us," says another.

This was exactly the good cop/bad cop routine the police had used on me.

Besides being compliant, I also showed signs of having made an "internalized" false confession. Sitting in that airless interrogation room in the *questura*, surrounded by people shouting at me during forty-three hours of questioning over five days, I got to the point, in the middle of the night, where I was no longer sure what the truth was. I started believing the story the police were telling me. They took me into a state where I was so fatigued and stressed that I started to wonder if I *had* witnessed Meredith's murder and just didn't remember it. I began questioning my own memory.

Kassin says that once suspects begin to distrust their own memory, they have almost no cognitive choice but to consider, possibly accept, and even mentally elaborate upon the interrogator's narrative of what happened. That's how beliefs are changed and false memories are formed.

That's what had happened to me.

I was so confused that my mind made up images to correspond with the scenario the police had concocted and thrust on me. For a brief time, I was brainwashed.

Three years after my "confession," I'd blocked out some of my interrogation. But the brain has ways of bringing up suppressed memories. My brain chooses flashbacks—sharp, painful flashes of memory that flicker, interrupting my conscious thoughts. My adrenaline responds as if it's happening in that moment. I remember the shouting, the figures of looming police officers, their hands touching me, the feeling of panic and of being surrounded, the incoherent images my mind made up to try to explain what could have happened to Meredith and to legitimize why the police were pressuring me.

This new knowledge didn't stop my nightmares or flashbacks, but I was so relieved to learn that what I'd been through wasn't unique to me. It had been catalogued! It had a name! As soon as I understood that what happened during my interrogation wasn't my fault, I started forgiving myself.

Kassin and others show that interrogations are intentionally designed to bewilder and deceive a suspect. Originally created to get highly trained, patriotic U.S. fighter pilots to sell out their country during the Korean War, one technique uses a tag team of investigators and tactics meant to induce exhaustion, agitation, and fear. It's especially potent on young, vulnerable witnesses like me. The method was designed not to elicit information but to plant it—specifically tailored to destroy an orderly thought process. After some hours, the subject gives the interrogators what they want—whether it's the truth or not.

In my case they'd put several interrogators in a room with me. For hours they yelled, screamed, kept me on edge. When they exhausted themselves, a fresh team replaced them. But I wasn't even allowed to leave to use the bathroom.

These were strategic measures, many of which are described in Kassin's report on police interrogation, "On the Psychology of Confessions: Does Innocence Put Innocents at Risk?" Reading it, I was flabbergasted to learn how by-the-book the police had been in their manipulation of me.

It had been the middle of the night. I'd already been questioned for hours at a time, days in a row. They tried to get me to contradict myself by homing in on what I'd done hour by hour, to confuse me, to cause me to lose track and get something wrong. They said I had no alibi. They lied, saying that Raffaele had told them I'd asked him to lie to the police. They wouldn't let me call my mom. They wouldn't let me leave the interrogation room. They were yelling at me in a language I didn't understand. They hit me and suggested that I had trauma-induced amnesia. They encouraged me to imagine what could have happened, encouraged me to "remember" the truth because they said I had to know the truth. They threatened to imprison me for thirty years and restrict me from seeing my family. At the time, I couldn't think of it as anything but terrifying and overwhelming.

That was exactly their point.

Sometimes I went over things I wish I'd done differently. Number one, I would have written to the Kerchers. I wanted to tell them how much I liked their daughter. How lovingly she spoke of her family. Tell them that her death was a heartbreak to so many.

Number two, I'd have written Patrick an apology. Naming him was unforgivable, and he didn't deserve it, but I wanted to say that it wasn't about him. I was pushed so hard that I'd have named anyone. I was sorry.

I didn't write then because Luciano and Carlo said not to con-

The haircut I got in protest after my conviction is evident
as guards escort me into the courthouse during my appeal in late 2010.
(Franco Origlia/Getty Images)

Rudy Guede testifying against Raffaele and me during our appeal. The appellate court reduced his thirty-year sentence to sixteen years. *(Alberto Pizzoli/AFP/Getty Images)*

Facing page: I gave my parents a small smile in the courtroom each day but behaved and dressed more somberly during my appeal than I had during my first trial. *(Mario Laporta/AFP/Getty Images)*

My lawyer Luciano Ghirga, greeting me in court, has always treated me like a daughter. *(Alberto Pizzoli/AFP/Getty Images)*

As the international media descended on Perugia to cover the verdict in Raffaele's and my appeal, TV cameras were allowed in during the court proceedings.

Days before the verdict in my appeal, I couldn't eat or sleep for fear of the outcome. *Below left:* Vice-Comandante Argirò of Capanne prison. *(Tiziana Fabi/AFP/Getty Images)*

Journalists throng the courthouse in Piazza Matteotti, awaiting our nighttime verdict. *(Tiziana Fabi/AFP/Getty Images)*

This is the first photo of our reunited family,
taken the day after I was freed.

I wasn't prepared to speak at the press conference after we landed
in Seattle and couldn't wait to embrace my family and enjoy my freedom.
(John Lok/The Seattle Times)

tact the Kerchers or Patrick. "They'll think it's a sympathy ploy," they said.

This made sense in the months after my arrest, but as my appeal approached, I had to set these wrongs right. I wrote letters to both Patrick and the Kerchers.

I wrote to Patrick first.

> *Dear Patrick,*
> *The explanation you've heard a number of times about my interrogation is true and I'm sure you understand well since you were arrested the same night without being told why.*
> *I feel guilty and sorry for my part in it.*

To the Kerchers, I wrote,

> *I'm sorry for your loss, and I'm sorry it's taken me so long to say so. I'm not the one who killed your daughter and sister. I'm a sister, too, and I can only attempt to imagine the extent of your grief. In the relatively brief time that Meredith was part of my life, she was always kind to me. I think about her every day.*

I showed the letters to Carlo. "It's not the right time," he said.

Disappointed and unsatisfied, I went back to my cell and came up with Plan B. I'd make a personal statement at the beginning of the trial. Unlike my declarations during the first trial, this one would be "spontaneous" in name only. I'd weave in Kassin's work to explain why I'd reacted to my interrogation as I had. At the same time, I'd speak directly to Patrick and the Kerchers.

I spent over a month writing drafts. Alone in my cell, I paced, muttering to myself as if I were speaking to the judges and jury.

As I honed my statement, I decided it would be stronger to speak from my heart, without Kassin's academic language. I'd tell the court about how I had been confused by the police and had lacked the courage to stand up to the authorities when they demanded that I name a murderer.

During the first trial, I believed my innocence would be obvious. It hadn't saved me, and I might never again have the chance to approach Patrick and the Kerchers. This time I was determined to help myself.

Chapter 32

December 11, 2010–
June 29, 2011

O ne must necessarily begin with the only truly certain, undisputed, objective fact: on November 2, 2007, a little after one P.M., in the house of Via della Pergola, Number Seven, in Perugia, the body of the British student Meredith Kercher was discovered."

Those were the opening words spoken at my appeal, by the assistant judge, Massimo Zanetti.

This trial looked much like our first one did. It was held two floors below ground in the Hall of Frescoes. The same people sat in the same places. The hundred-plus attending journalists called out the same questions: "Do you feel confident?" "Do you have anything to say?" One yelled, "Nice shoes!"

Everything may be as it was, I thought. *Except me*. I was conscious of how everything I did, said, or wore, every face I made could contribute to the outcome, good or bad.

I stepped over the threshold feeling full of dread. *The last time I was in this room I had to be carried out.*

Madison, Mom, and Chris were sitting together. I gave them a careful, barely-there smile. Anything bigger would beg another "Smiling Amanda" headline.

Clothing seemed a poor criterion for determining if I was going to spend two and a half decades or more in jail, but I couldn't afford the distraction of casual clothes. Rocco and Corrado had given Laura money to buy me appropriate court clothes. She turned out to be an excellent personal shopper.

My champagne-colored blouse and black pants told the judges and jury that I respected them and the law.

Raffaele also had a new look—though I don't know if it was meant to appeal to the jury or just himself. He had beefed up and buzzed his hair during the fifty-three weeks since I'd seen him. I'd lost enough weight that the escort guards who gripped my arms said, "*Devi mangiare di più!*"—"You need to eat more!"

The judge's opening statement gave us hope that the court wanted a trial grounded in facts, not theories. *Will we finally get a fair trial? Will the judges and jury finally listen to what we have to say?*

I stood to deliver my declaration, the one I'd worked on for weeks. Speaking in Italian, without an interpreter, I sensed my voice quavering, my hands trembling:

> *"I was wrong to think that there are right or wrong places and moments to say important things. Important things have to be said, no matter what . . . To Meredith's family and loved ones, I want to say that I'm so sorry that Meredith is not here anymore. I can't know how you feel, but I, too, have little sisters, and the idea of their suffering and infinite loss terrifies me."*

Since the Kerchers weren't there, I addressed my comments to the jurors. And though I was desperate not to cloud my message by crying, I choked up at the mention of my three siblings.

> *"It's incomprehensible, unacceptable, what you're going through, and what Meredith underwent. I'm sorry all this happened to you and that you'll never have her near you again, where she should be. It's not right and it never will be. You're not alone when you're thinking of her, because I'm thinking of you, I also remember Meredith, and my heart aches for all of you. Meredith was kind, intelligent, nice, and always accommodating. She was the one who invited me to see Perugia with her, as a friend. I'm grateful and honored to have been able to be in her company and to have been able to know her."*

I turned toward Patrick, whose lawyer blocked my sight line of him.

> *"Patrick? I don't see you. But I'm sorry. I'm sorry, because I didn't want to wrong you. I was very naïve and not remotely courageous, because I should have been able to endure the pressure that pushed me to wrong you. I didn't want to contribute to all that you suffered. You know what it means to have unjust accusations imposed on your skin. You didn't deserve what you went through. I hope you'll succeed in finding your peace."*

Next I spoke to the court.

> *"I never expected to find myself here, condemned for a crime I didn't commit. In these three years I've learned your language and I've seen how the judicial procedure works, but I've never gotten used to this broken life. I still don't know how to face all this if*

not by just being myself, who I've always been, in spite of the suffocating awkwardness . . . Meredith's death was a terrible shock for me. She was my new friend, a reference point for me here in Perugia. But she was killed. Because I felt an affinity toward her, suddenly, in her death, I recognized my own vulnerability. I clung above all to Raffaele, who was a source of reassurance, consolation, accommodation, and love for me. I also trusted the authorities carrying out the investigation, because I wanted to help render justice for Meredith. It was another shock to find myself accused and arrested. I needed a lot of time to accept that reality, of being accused, and redefined unjustly. I was in prison, my photo was everywhere. Insidious, unjust, nasty gossip about my private life circulated about me. Living through this experience has been unacceptable for me. I trust above all the hope that everything will be worked out as it should be, and that this enormous error about me will be recognized, and that every day that I spend in a cell and in court is one day nearer to my freedom. This is my consolation, in the darkness, that lets me live without despairing, doing my best to continue my life as I always have, in contact with my friends and family, dreaming about the future. Now, I am unjustly condemned, and more aware than ever of this hard and undeserved reality. I still hope for justice, and dream about a future. Even if this experience of three years weighs me down with anguish and fear, here I am, in front of you, more intimidated than ever, not because I'm afraid or could ever be afraid of the truth, but because I have already seen justice go wrong. The truth about me and Raffaele is not yet recognized, and we are paying with our lives for a crime that we did not commit. He and I deserve freedom, like everyone in this courtroom today. We don't deserve the three years that we've already paid, and we certainly don't deserve more. I am innocent.

Raffaele is innocent. We did not kill Meredith. I beg you to truly consider that an enormous mistake has been made in regard to us. No justice is rendered to Meredith or her loved ones by taking our lives away and making us pay for something we didn't do. I am not the person that the prosecution says I am, not at all. According to them, I'm a dangerous, diabolical, jealous, uncaring and violent girl. The people who know me are witnesses to my personality. My past, I mean my real past, not the one talked about in the tabloids, proves that I've always been like this, like I really am, and if all this is not enough, I invite you to ask the people who have been guarding me for three years. Ask them if I have ever been violent, aggressive, or uncaring before the suffering that is part of the broken lives in prison. Because I assure you that I'm not like that. I assure you that I have never resembled the images painted by the prosecution. How is it possible that I could be capable of achieving the kind of violence that Meredith suffered? How is it possible that I could throw myself like that at the opportunity to hurt one of my friends, for the sake of violence, as though it were more important and more natural than all my education, all my values, all my dreams, and my whole life? All this is not possible. That girl is not me. I am the girl that I have always shown myself to be and have always been. I repeat that I also am asking for justice. Raffaele and I are innocent, and we want to live our lives in freedom. We are not responsible for Meredith's death, and, I repeat, no justice is accomplished by taking our lives away. Thank you."

I had hoped that I could have talked to both Patrick and the Kerchers privately. Now that seemed extraordinarily unlikely. Still, it felt good to say what needed saying. *Finally, I've done the right thing.*

It was the longest, most emotionally exhausting seventeen minutes of my life.

And it might have been my only chance if the court ruled against our request for an independent review of Raffaele's kitchen knife and the bra clasp and bringing in new witnesses. My lawyers believed there was a good chance they would grant it, but "if not," Luciano had said, "you need to be strong. We'll make our case anyway."

My declaration left me feeling cleansed and relieved. I didn't expect to change minds instantly—and I didn't. Chris, Mom, and Madison told me later that the Kerchers' lawyer, Francesco Maresca, had left the room at my first mention of Meredith's family. "She bores me," the London *Guardian* reported him saying. "Her speech lacked substance, was designed to impress the court and was not genuine."

Maresca cared more about seeing me convicted than finding justice for Meredith. He always spoke of me as if I were a monster who must pay for Meredith's death with my life.

I had to hope that the judges and jury—five women and one man—didn't feel that way about me.

I knew the first court hadn't convicted us solely on the forensic evidence, but I couldn't imagine how Raffaele's and my team could defend us if the appeals judge, Claudio Pratillo Hellmann, and his second, Massimo Zanetti, squashed our demands a second time.

That's what the prosecution was pushing for. A review is "useless," they said. "This court has all the elements to be able to come to a decision."

Since court hearings were held only on Saturdays, an excruciatingly slow week would have to pass before we'd know Judge Hellmann's mind. While we waited, Italy's highest court signed the final paperwork on Rudy Guede's verdict, approving his reduced sixteen-year sentence in the belief that he had not acted alone. Could that news influence Judge Hellmann's decision? By pursuing our trial, he might seem to be contradicting the Supreme Court and make Italy look foolish.

"Do you think that will hurt us?" I asked Chris and Madison at their next visit to Capanne.

"All I know is that you'd better keep building up your iron *cojones*," Chris answered, trying to make me laugh.

The following Saturday, when the court retired to chambers to decide, I tried to calm my jangled nerves. Unfortunately, cameras were allowed in during breaks. The photographers bent themselves sideways trying to zoom in on me.

During the hour-and-twenty-minute wait, I occasionally turned around long enough to meet my mom's or Madison's eyes. Mom smiled nervously at me. "Courage," my lawyers reminded me. "Courage."

When Judge Hellmann came out to announce his decision, I held my breath and squeezed Luciano's hand, instinctively ducking my head to avoid a painful blow.

"I'm convinced the case is complex enough to warrant a review in the name of 'reasonable doubt,'" Judge Hellmann told the rapt courtroom. "If it is not possible to check the identity of the DNA, we will check on the reliability of the original tests."

Maria Del Grosso, from Carlo's office, patted me excitedly on the shoulder.

I hadn't wanted to admit to my lawyers or to myself how petrified I'd been. Only when the result came back did I realize how much fear I had had pent up. I brushed away tears. *We might finally have a real chance to defend ourselves.*

Still, I was wary. The judge in the previous trial had granted

our request for data and then sided with the prosecution's inter-
pretation.

———————

After that, we were back to waiting again. The independent
experts, Dr. Carla Vecchiotti and Dr. Stefano Conti, forensic medicine professors at Rome's university, La Sapienza, were
sworn in, and Judge Hellmann charged them with figuring out
whether a new analysis of the DNA on the knife and bra clasp
was possible. If not, he wanted to know if the original results of
the prosecution's forensic expert were reliable: Were the interpretations of the genetic profiles correct? Had there been risk of
contamination? The experts were given three months from the
day the prosecution turned over the evidence.

While the experts were working, Judge Hellmann moved
ahead with the new testimony he'd granted. During the first trial,
Prosecutor Mignini had called the witness Antonio Curatolo, a
homeless man referred to as "the stepping-stone leading us up
to the murder." Curatolo had testified that he'd seen Raffaele
and me arguing on the basketball court in Piazza Grimana. It
was key evidence in our conviction, because it contradicted our
alibi that we'd never left Raffaele's apartment. But it had been
left unclear which night Curatolo was describing—Halloween
or November 1?

Curatolo was recalled as a witness, but he came under different circumstances. The onetime homeless man was now in
prison himself, on drug charges. He arrived in the courtroom
flanked by two guards, just as Raffaele and I were. As he took the
stand he said that the night he saw Raffaele and me, "there were
a lot of young people in costume" joking around. "There were

other people who were messing around. It was a holiday." Buses were there to take young people to discos outside town.

Raffaele's lawyer, Giulia Bongiorno, asked, "So you're saying, the night that you saw Raffaele and Amanda, there were people wearing masks and in buses?"

To the defense it was obvious this description matched Halloween, not November 1, a religious holiday, when the clubs were closed and no buses had been hired for the night.

Curatolo had also testified that the day after seeing Raffaele and me in Piazza Grimana, he saw Carabinieri and men dressed in white—"Martians," he called the Polizia Scientifica in their anticontamination suits. Since the Polizia Scientifica came to the villa on November 2, this meant that Curatolo must have seen Raffaele and me on November 1.

"So the very next day that you saw Raffaele and Amanda, there were police officers and people in white suits at the villa?" Prosecutor Mignini asked.

"I am as certain of that as I am sure that I am sitting in this chair now," Curatolo told the appeals court.

The prosecutor and civil lawyers insisted that some of the discos were open on November 1, 2007, and that there were buses in Piazza Grimana.

Fortunately the court allowed our defense teams to call new witnesses, the managers of Perugia's large discos, to the stand. Halloween, they said, is "the biggest night of the year." A witness from Red Zone, where I'd gone with Meredith and the guys downstairs, added, "There were no buses" on November 1.

"I'm certain," she said, "because discos focus on Halloween, which is a big draw. It's like New Year's Eve."

Under the judges' questioning, Curatolo talked about his personal history: "I was an anarchist, then I read the Bible and became a Christian anarchist," he said. He confirmed that he was now in prison, adding, "I haven't quite understood why yet." Asked if he'd used heroin in 2007, he answered, "I have always used drugs. I want to clarify that heroin is not a hallucinogen."

I'd prepared notes for a statement but abandoned them. Curatolo was doing a good enough job muddling his witness statement and making a fool of the prosecution, who still claimed him to be a "decisive"—aka "super"—witness.

That night I was able to call home. Mom, Chris, and Madison were in Perugia, but I called Seattle each week to talk to my dad and stepmom, my sisters, Oma, my aunts, uncles, and cousins, and assorted college friends. After a chorus of "hellos," everyone asked, "How'd it go?"

They'd seen the news but wanted to hear firsthand from me.

"Curatolo didn't know what he was talking about, poor guy. If my life didn't depend on his being wrong, I'd just feel bad for him," I reported.

"The broadcasts here are saying that he's a confused drug addict!" someone cried.

It was ironic that I learned from my family in Seattle what the journalists in the courtroom were thinking. "The media are really figuring it out this time," my family reassured me. "It's going to be okay."

The media, yes. *But what about the judges and jury?* I wondered. Curatolo hadn't been convincing in the first trial, either, but his testimony had contributed to our conviction.

Everything hung on the independent review.

On the phone, during prison visits, in my letters, buoyed by the people I loved, I allowed my confidence to gain over my doubt and fear. But I spent most of my time on my own. I was a prisoner. Until someone unlocked my cell and told me I was being let out of this suffocating hell, I could only hope. But for the past three-plus years, hope had let me down.

————————

Back in January, when Judge Hellmann swore in the independent experts, Conti and Vecchiotti, he gave them ninety days to analyze the forensic data and submit their conclusions to the court. The clock would start when the prosecution handed over the evidence.

Before the first trial, the defense began requesting forensic data from the prosecution in the fall of 2008, but DNA analyst Patrizia Stefanoni dodged court orders from two different judges. She gave the defense some of, but never all, the information. Now it was Conti and Vecchiotti's turn to try to get the raw data that Stefanoni had interpreted to draw conclusions about the genetic profiles on the knife and the bra clasp. Stefanoni continued to argue that the information was unnecessary. Not until May 11, under additional orders from Judge Hellmann, did she finally comply.

Now the independent experts needed more time. My lawyers said judges always grant leeway when experts ask. Before the court withdrew to decide whether to approve the delay, I made a statement. "I've spent more than three and a half years in prison as an innocent person," I told the court. "It's both frustrating and mentally exhausting. I don't want to remain in prison, unjustly, for the rest of my life. I recall the beginning of this whole thing, when I was free. I think of how young I was then, how I didn't understand anything. But nothing is more important than finding the truth after so many prejudices and mistakes. I ask the court to grant the extra time, so that the experts may complete a thorough analysis. Thank you."

I knew in advance from my lawyers that the independent experts were going to ask for more time, so I tried to imagine what that meant to me. An extra forty-five days meant the appeal would likely last through the summer. I dealt with the wait by reminding myself that another month and a half was tolerable as long as it meant that I wouldn't spend my life in prison. I thought it was critical to get this idea across to the judges and jury.

Judge Hellmann and the court retired to chambers, only to return shortly afterward, agreeing to give the forensics experts until June 30 to deliver their report.

About a week later came a reminder not to get overconfident. Police Holiday is an annual event in Umbria, when awards are traditionally handed out for outstanding police work. Perugia is the region's capital.

When Luciano came to Capanne for our weekly Wednesday meeting, he told me that a special award had been given to officers in the Squadra Mobile for its work on Meredith's murder investigation.

The citation read: "To recognize elevated professional capabilities, investigative acumen, and an uncommon operative determination. They conducted a complex investigation that concluded in the arrest of the authors of the murder of the British student that had taken place in the historic center of Perugia."

Four of the sixteen police officers receiving the Police Holiday award were named in the police's slander charge against me.

They included Vice Superintendent Marco Chiacchiera, whose "investigative instinct" led him to randomly select Raffaele's kitchen knife from the drawer as the murder weapon; Substitute Commissioner and Homicide Chief Monica Napoleoni; and Chief Inspector Rita Ficarra.

The news infuriated me. I knew it was just another face-saving ploy. How could they commend the officer who had hit me during my interrogation and those who had done so much wrong?

But I wasn't surprised. It was completely in line with the prosecution's tactics to discredit my supporters and me. Mignini had charged my parents with slander for an interview they gave to a British newspaper in which they told the story of my being slapped during the interrogation. He was the one who had charged me with slandering the police.

Journalists had started cataloguing the mistakes in the investigation. Most were made by the Squadra Mobile, which relied on intuitive judgments instead of evidence. It was the same haste-makes-waste argument that Luciano and Carlo had made in their closing arguments for the first trial, focusing on the pressure police had been under to arrest a suspect and how that had led to police errors.

British journalist Bob Graham interviewed Mignini for an article in *The Sun* that came out on Police Holiday. Mignini confided in Graham that he chose the parts of my interrogation that suited his purposes. He also said that my interpreter at the *questura* that night was "more investigator than translator." When Graham asked the prosecutor why there was no evidence of me in Meredith's bedroom, Mignini told him, "Amanda might theoretically have instigated the murder while even staying in the other room."

Mistakes or not, the police's message was crystalline: We're not backing down now.

———————

My focus was on the courtroom, where there was more new testimony to hear.

Mario Alessi was a brick mason given a life sentence for murdering an infant boy in 2006. He was in the same prison as Rudy Guede, and had written to Raffaele's lawyers that he had information for our defense:

Alessi said he went outside for exercise with other prisoners, including Rudy Guede, on November 9, 2009. "Guede told me he wanted to ask me for some confidential advice," Alessi said in his court deposition. "There wasn't a day that Guede and I didn't spend time together . . .

"In this context, on November 9, 2009, Guede told me that in the following days, and in particular on November 18, 2009, he had his appeal and he was reflecting over whether to . . . tell the truth about Meredith Kercher's murder. In particular, he asked me what the consequences could be to his position if he gave statements that reconstructed a different truth about what happened the night of the murder.

"I responded that I wasn't a lawyer, and I didn't know what to say, but that I believed it would be useful to tell the truth. So he confided in me, describing what happened the night of the murder."

Guede told Alessi that he and a friend had run into Meredith in a bar a few days before the murder. On the night of November 1, Alessi said, the two men surprised Meredith at the villa and, "in an explicit manner," asked her to have a threesome.

Alessi said that Meredith "rejected the request. She even got up and ordered Guede and his friend to leave the house. At this point Guede asked where the bathroom was, and he stayed in the bathroom for a little while, ten to fifteen minutes at most. Immediately after, reentering the room, he found a scene that was completely different—that is, Kercher was lying with her back to the floor and his friend held her by the arms. Rudy straddled her and started to masturbate. While Guede told me these things, he was upset and tears came to his eyes . . .

"The second part of his secret came out while we were in our respective cells . . . at a certain point he and his friend changed positions, in the sense that his friend attempted to have oral sex with Meredith while Guede was behind. He specified in particular that his friend was in front of Meredith, who was on her knees, while Guede was behind Meredith, with his knee on her back. Kercher tried to wriggle out . . .

"Kercher tried to get away, and at this point Guede's friend took a knife with an ivory-colored handle out of his pocket. While Kercher tried to get away, turning around, she was wounded by the blade. At this point, seeing as she began to bleed, Guede, finding his hands covered in blood, let her go. While Guede tried to staunch the wound with clothes, his friend reprimanded him, saying, 'Let's finish her. If not, this whore will have us rot in prison.' At this point, his friend killed her, stabbing her various times while Guede gathered clothes to staunch the wounds. Then, realizing that she wasn't breathing anymore, he left."

After the murder, Guede went to a club and met his murderer friend, who gave him money and told him to flee Italy.

Alessi said, "Guede, at my questioning, responded that he couldn't say whether or not it was the money that was stolen from Kercher. I also asked Guede how he could explain the broken window and the rock that was found in Kercher's house, but Guede responded that while he was in the house he hadn't heard a sound and didn't know anything about that window."

Alessi said that when he suggested that Guede tell the truth, "because there were two innocent people in prison . . . Guede responded that he certainly wasn't the one to put those two in the middle of everything but rather the prosecution . . .

"I can also refer to an episode in Cell No. 11 in the presence of Antonio De Cesare, Ciprian Trinca, and Rudy Guede"—all prisoners.

"We were playing cards and, once again, in the course of a television program, Meredith's murder was brought up, and at that point Luca Maori, Raffaele Sollecito's defense counsel, was being interviewed. Guede made a comment against Sollecito . . ."

Guede said that since he didn't have the same opportunity to defend himself as Raffaele, he was the victim.

Listening to Alessi testify, I felt frozen in my chair, my limbs numb. Alessi was a calm, direct, convincing speaker. *Is this possibly what happened the night of November 1? Is this the horror that Meredith experienced?* For three and a half years, I'd tried to imagine Meredith's murder and had to push it out of my mind. When the prosecutor had put Raffaele and me into the scene, it hadn't bothered me nearly this much. We weren't there, so Meredith's murder couldn't possibly have unfolded the way Mignini described. His story was so far-fetched, and it was so painful to hear myself described in bloodthirsty terms, that I couldn't help but focus on the verbal attack on me rather than the physical attack on Meredith.

Alessi's story, however, sickened me when I heard it and haunted me long after. I knew it was only hearsay and that even though two of Guede's other prisonmates corroborated it, it couldn't be used as direct evidence. Real or not, it forced me to focus on the torture that Meredith was put through. And it opened up a question I'd never seriously considered and could barely handle: Had there been someone with Guede?

My lawyers once told me that investigators had found unidentified DNA at the crime scene, but I'd never dwelled on it. The prosecution had never presented it. Wouldn't there have been signs of another person in the room and on Meredith's body? I didn't know. This is what I *was* sure of: Guede was there, Guede lied about us, Guede tried to escape his responsibility for the crime.

Guede would have to confess.

I desperately hoped he'd be honest when he took the witness stand. With the Supreme Court's seal on his conviction, his sentence couldn't be extended no matter how he incriminated himself. Since he truly had nothing to lose, I thought he might admit his crimes—and the fact that Raffaele and I weren't there that night.

I planned to make a spontaneous declaration directly to him, either challenging him to tell the truth or thanking him for doing so. For a week I thought about what I should say, pacing in my cell as I tried out different words. I'd written Guede a letter I'd never sent. I wove that into my declaration just as I'd done with my statement to Patrick and the Kerchers.

In the meantime, I was agitated. I had no reason to expect that Guede would admit what had happened—anyone who can kill is already lacking a conscience. Even if Guede acknowledged Raffaele's and my innocence, it still wouldn't be enough on its own to free us—his statements were compromised since he'd lied before and wasn't impartial. But it would be a huge step in the right direction—and an even bigger comfort to me.

Taking the witness stand, Guede said he wouldn't speak about the murder, that he was there only to respond to his former prisonmate's accusations. Mignini read a letter to the court that Guede had supposedly written to his lawyers after Alessi's claims surfaced. I found it so unsettling that I could hardly listen. The letter didn't remotely correspond with Guede's education—he wasn't a model student and, in fact, had dropped out of school. The language was sophisticated. Calling Alessi "a vulgar being with a foul conscience," the letter condemned his "blasphemous insinuations." It ended with a comment on "the horrible murder of a splendid and wonderful girl by Raffaele Sollecito and Amanda Knox."

When Carlo tried to pin him down, Guede told an attentive court, "It's not up to me to say who killed Meredith. I've always said who was in the house that damned night."

I couldn't contain my anger another second. I had to denounce his lies. Just as Guede was about to be dismissed, I asked to make a spontaneous declaration. Judge Hellmann said I'd have to wait until Guede had left the courtroom.

I felt cheated as I watched him walk out of the courtroom in handcuffs. I was disgusted. I'd truly hoped and believed that Guede would do the right thing, because, damn it, he was human. How could he not, ultimately? As the double doors closed, I quickly reorganized the statement I'd prepared. "I just want to say that the only time Rudy, Raffaele, and I have been together in the same place is in court," I said. "I'm shocked and anguished by his testimony. He knows we weren't there."

Then I sat down, crying.

Chapter 33

———————

June 29, 2011

W/*hat if?*

Twenty-four hours before the court-appointed experts were to present their findings on the DNA, only two words were going through my mind. *What if?* What if their review somehow—impossibly—confirmed Meredith's DNA on the knife blade? What if they found that the bra clasp couldn't have been contaminated?

Or what if the experts risked telling the truth and sided with the defense?

I knew the prosecution's DNA testing was flawed. But so little had gone right in this case, why would this go right?

Science was on our side. The knife blade had tested negative for blood, and there was a high likelihood that the bra clasp had been contaminated while it sat on the floor for six weeks. But I had no faith in facts anymore. They hadn't saved me before.

It was terrifying to hope—and impossible not to.

Over the summer, there were moments when I could escape the pressure and just be me—a twenty-three-year-old girl. Most afternoons Don Saulo called me down to his office, and we spent an hour together. It was precious time for me.

I'd gotten used to telling him everything on my mind. I appreciated his intelligence, compassion, and intuition. Don Saulo had lived a sheltered life, but his empathy was strong and unreserved.

I talked to him about my family and friends, my ideas and doubts.

We also talked about music. Don Saulo had invited me to play guitar at Saturday afternoon Mass. Now he was teaching me basic music theory and how to play an old electronic keyboard he had in his office. We'd listen to a song on his portable CD player, and then we'd learn it together on the guitar or piano. I drew a paper keyboard so I could practice in my cell at night—ear buds in, playing the silent chord progressions.

But I was too nervous to play the day before the experts' announcement. Don Saulo sat across from me holding my cold hands in his warm ones as I plodded through my "what if" possibilities.

"No matter what happens," he said, "live your life to its fullest."

I looked down. "Yeah, no matter what happens, I can only make the best of it."

When the hour ended, I went back to my cell as dejected as ever. My current cellmate, Irina, was sitting on her bed smiling.

"What's up with you?" I asked.

"Oh, just a little news," she purred.

She must know something about the review. I imagined the news must be good. *She has that maddening smile. But what if she's mocking me?*

"Everyone's saying it," she burst out. "You're going home!"

"The report is out?!" I screamed. "It's okay?! I'm going home?! Says who?"

"The TV! The news. The forensic report came back! You're cleared! They said you'll be freed!"

I had to hear the words myself. I went to the TV, madly changing channels until I found the news. *"Svolta Giudiziaria"*— "Judicial Turning Point," the headline read, behind an announcer who was talking about my case. The crawl at the bottom read: "DNA damning Knox and Sollecito deemed unreliable by court-appointed experts. New hope arises for the defendants."

Suddenly my heart was filling my whole chest. I couldn't breathe. From the moment I'd been arrested, I'd never heard good news about my case on TV. *It's finally happening!* I jumped up and down and spun around in a little dance, whooping and yelling, "I can't believe it! I can't believe it!

Stefanoni, the prosecution's forensic DNA expert, was finally recognized as being wrong.

I was crying, my face flushed and hot.

Irina opened her arms, and I squeezed her as tightly as I could.

"Ah, Amanda!" she said, laughing. "You're going to pop my implants!"

It was the first time in three and a half years at Capanne that I could truly jump for joy.

I remembered Don Saulo. I'd just burdened him with my angst. *I have to tell him!* I ran to the bars of my door. *"Assistente!"* I bobbed on my toes while I waited for the *agente* who'd opened my cell five minutes before. I knew asking her to unlock the door again was a no-no. *I don't care! I have to tell Don Saulo!*

The *agente* approached the door looking bored. "What is it, Kuh-nox?" she asked sourly.

"I know I was just down to see Don Saulo," I said breathlessly, "but I have to go back down—just for a second. I have to tell him the news. The forensic report came back, and everything is okay. I have to tell him, because I didn't know before. Okay?"

My hands were on the bars and I was leaning into the *cancello* as though willing it open.

The *agente* eyed me with confusion. "You want to see Don Saulo again?"

"Please. Just for a second!"

She looked perplexed as she turned the key and swung the barred door open. I rushed out, jogging down the hallway, even though it wasn't permitted. I called back to the guard, "I'll be back in a second!"

Don Saulo was in the chapel leading a group of prisoners in Bible study. I rushed inside, beaming, and hugged him. "It came out!" I whispered. "It's good!"

When I pulled back, I saw that he'd teared up. The women stared. They'd seen Don Saulo cry plenty of times, but they'd never seen me excited. "What happened?" one asked.

"The forensic report came out. It supports the defense," I said. "I might actually be freed!"

"You see! There is God! There is God!" exclaimed Tessy, one of the Nigerian women I'd help write letters to her family. She jumped up and hugged me. So did Beauty, another Nigerian.

I said good-bye and went back upstairs. At the gate to the hallway, the *agente* saw me and glared.

"I'm really sorry," I said. "I had to tell Don Saulo the news about my case."

"You could have told him tomorrow," she grumbled.

The rest of the evening I flipped channels, watching news report after news report, wanting to hear the words again and again—"*Svolta giudiziaria. Nuova speranza per Amanda e Raffaele*"—"Judicial turning point. New hope for Amanda and Raffaele."

Chapter 34

June 30–October 2, 2011

T he next morning I arrived at the Hall of Frescoes with a lighter heart. The journalists called out, "Amanda, what do you think of the new findings?" "Are you excited?" "Do you think you're going home?"

I didn't answer, but I liked the tone of these new questions.

I could see my mom trying to suppress her glee. When I got to the table, Carlo squeezed my hand. Raffaele nodded and smiled. We were all trying to contain ourselves. We weren't in the clear yet, but we were closer than we'd ever been. And I think we all had the deep-seated fear that somehow the prosecution would flip the findings and convince the judges and jury that the old report was the right report. I knew they'd try. They'd been publicly embarrassed.

This time the trial was going our way. I was delighted—I hoped it wasn't obvious—when the experts criticized the Polizia Scientifica's procedures. My DNA was on the knife handle, but the DNA trace on the blade was "unreliable," because Patrizia Stefanoni had ignored international protocol in testing such a tiny amount. It could have come from contamination, they said.

Professor Stefano Conti showed the video of the Polizia Scientifica collecting evidence when they returned to the villa six weeks after Meredith was killed. The professor zoomed in on the dirty latex gloves the investigators wore. The police's own recording showed them passing the bra clasp back and forth and then putting it back on the floor to photograph as evidence. "There are a number of circumstances that don't follow protocol or proper procedure," Conti said in something of an understatement.

By the video's end, he'd identified more than fifty mistakes the forensics team had made, including waiting six weeks to collect the evidence, using the wrong type of bags to collect evidence, wearing gloves dotted with blood and dirt, and picking up Meredith's bra and underwear and touching her body barehanded.

"Today was a profound, clear, and unequivocal analysis of the DNA on the bra clasp," said Raffaele's attorney Giulia Bongiorno. "DNA on the bra clasp attributed to Raffaele Sollecito was the only evidence on which he was convicted. This so-called evidence has fallen apart."

As the weeks went by, I was starting to have faith that this judge wouldn't overlook the mistakes the police had made.

As expected, the prosecution and the civil attorneys tried to delegitimize the experts by saying they were biased in favor of the defense and complaining that neither expert was qualified.

They were reaching.

The co-prosecutor, Manuela Comodi, said Conti and Vecchiotti were lying. Show us the exact moment when the bra clasp was contaminated, she said. If we couldn't prove it was contaminated, we couldn't claim it.

Vecchiotti and Conti's response: Following protocol is the

way a forensic scientist proves that contamination doesn't happen. The forensics team picked up the bra clasp that was found in a different part of the room, put it down, photographed it, and picked it up again, and you're saying there wasn't a high likelihood that it was contaminated?

You can't prove that the glove touching the bra clasp was contaminated, Comodi told the experts.

Conti and Vecchiotti said, "We have a picture of the glove. You can see the dirt."

The prosecution said you have to prove that the glove had Raffaele's DNA on it.

Conti and Vecchiotti's final words on the subject: No, we don't. It's enough to show that the glove was dirty and that the bra clasp was moved from one place to another, that it wasn't picked up for six weeks—that protocol was violated.

That day, July 30, was the last hearing before the August break. Judge Hellmann announced that he wanted the court to return to session on September 5. The co-prosecutor objected. "I was hoping to still be on vacation with my daughter then," she said.

On vacation with your daughter! I screamed in my head. *I wish I could be on vacation with* my *mother! You're worried about extending your vacation and you don't care that I've missed out on almost four years of my life!*

Judge Hellmann set the next hearing for September 5.

I didn't know when the verdict would come, but the closer we got, the more nervous I felt. I couldn't eat, my hair was again falling out in clumps, I was covered in hives, and my hands shook involuntarily. I often burst out crying. Mainly I couldn't relate to the uninhibited enthusiasm of my family, friends, and supporters. When Corrado visited in August, he asked, "Why are you so worried, Amanda? Everything's going to be fine. You'll see. Just relax."

I couldn't even draw a full breath.

The closest I came to unwinding was the time I spent playing music and talking with Don Saulo. The weather was too hot to walk in the afternoons, too hot to move during the day, almost too hot to think. I wrote lots of letters to James and others in Seattle, and to Laura in Naples. I read. I daydreamed about the four possibilities that awaited me when Judge Hellmann read out my verdict. Life imprisonment? Twenty-six years? A lower sentence? Acquittal? I broke my own rule and counted the days until September 5. I knew I shouldn't. It made the thirty-seven days between court dates crawl by.

When September finally arrived, being back in the courtroom helped me regain a tiny bit of control over my hypernervousness. It meant that things were happening again. It was better to focus on the momentum than the waiting.

The prosecution hired two other forensic experts to testify that contamination can be said to occur only if you can prove precisely where, when, and how.

One of our DNA experts, Sarah Gino, emphasized that Patrizia Stefanoni had been withholding data from the very beginning.

The defense's next expert, Carlo Torre, testified that the police's DNA testers had found no blood on Raffaele's kitchen knife. What the independent experts had found were traces of potato starch. If the knife had been cleaned with bleach, as the prosecution claimed, the starch wouldn't be there—and bleach wouldn't have entirely diluted the blood, if blood had ever been on the knife in the first place.

The prosecution asked for a new, independent review of the knife, but Judge Hellmann rejected the request. Instead he announced the schedule for closing arguments and the verdict—October 3.

Then there was yet another short break before closing arguments began.

In a fit of optimism, I decided what belongings I'd leave behind if I were acquitted. I didn't want the jeans and sweatshirts that I associated with prison or any of the day-to-day stuff I needed to exist—my camping stove, pots and pans, pens, paper, markers. I gave Chris books each time he came to visit. Over the weeks, he took away twelve boxes, each holding twenty to thirty books.

Packing made me nervous. I'd done this before and then had had to return to prison. It was embarrassing to sort my things in front of guards and other prisoners who probably thought it was futile. Some people were excited for me; others pulled away. Guards and prisoners kept telling me, "Promise you'll write to us. Promise you'll remember us."

I'd stay in touch with Don Saulo and Laura, but I didn't want to take the prison with me.

If my hopes were finally to come true, I'd be prepared. My belongings sat in a canvas bag in my cell. But I kept my pictures of family and friends out. I needed to look at them in my lonely moments—and I'd really need them close if things didn't go well in the end.

———

Closing arguments began on September 23 with Perugia's chief prosecutor, Giancarlo Costagliola, and Mignini insisting, "All clues converge toward the only possible result." The men asked the jury to ignore the hype in the media that favored Raffaele's and my acquittal, to uphold our conviction, and to keep the Kerchers in mind. Mignini said, "If you want, go ahead and believe that Rudy Guede is the only one, but we don't believe in fairy tales, and neither does the court."

I'd steeled myself for his detailed description of what I would have said to Meredith and how I'd killed her, but it still hurt. Every word jabbed me like a sharp stick.

Mignini added that as further evidence of my guilt, I was "ready to flee Italy" if I were acquitted.

He was not quite right. After four years of wrongful imprisonment, I'd kayak home if I had to. But if I were acquitted, my leaving was hardly "fleeing."

Accusing the defense teams of slandering Patrizia Stefanoni's forensic scientists, Mignini quoted the Nazi propaganda minister Joseph Goebbels, who famously said, "Slander, slander, something will always stick."

Manuela Comodi tore apart the independent experts' testimony. "They betrayed your trust with false facts," she said. "Their whole manner was aggressive when they should have been impartial."

She added, referring to Raffaele and me, "They are young, but Meredith was also. They are young, but they killed. They killed for nothing, and it is for this reason that they must be condemned to the maximum sentence, which, luckily, in Italy isn't the death penalty."

Carlo Pacelli, Patrick's lawyer, again emphasized that I was a "sorceress of deceit."

Intentionally or not, Francesco Maresca, the Kerchers' lawyer, ended with a shock tactic. Although the Kerchers had asked that no pictures of Meredith's naked, wounded body be shown without clearing the courtroom of reporters, Maresca projected the images on a screen. He said he wanted to show how Meredith had suffered, so the court wouldn't let us off on a "technicality."

Though they couldn't afford the airfare to attend the appeal, Meredith's mother and sister would be in Perugia for the verdict, he said. "They will look you in the eye . . . and with their look they will ask you to confirm the earlier sentence."

It was painful to hear the prosecution and civil parties suggest that justice could be rendered for Meredith and her parents only by putting us in prison for life. *That's not justice! Please don't confuse the two!* And justice had already been denied by the prosecution itself, when they let Guede get off with a lesser sentence than he merited.

Raffaele's lawyer Giulia Bongiorno brought up another way that justice had not been done. She spoke of the phenomenon of false confession, saying, "This is what happened to Amanda Knox."

She compared me to Jessica Rabbit in the movie *Who Framed Roger Rabbit?*: "I'm not bad. I'm just drawn that way."

Luciano, who referred to me as "this young friend," said, "Amanda isn't terrified. Her heart is full of hope. She hopes to go back home. I wish her that," he said. "I feel I'm going to cry . . . She is so brave, Amanda."

Not knowing what would happen was driving me mad.

I wanted to go to James's senior guitar recital at UW; to be at my twin cousins, Izzy and Nick's, sixth birthday party; and to see my little sister Delaney graduate from middle school. I wanted to go outside whenever I wanted, to feel the grass, to eat sushi.

In my journals I'd draw lines down the pages. On one side, the things I'd do if I got out now. On the other, things I'd do if I got out when I was forty-six. On the left, I wrote:

Move into an apartment with Madison.
Graduate from UW.
Visit Laura in Ecuador.
Write.
Become fluent in German so I can talk with Oma.
Go camping and hiking with my family.
Pay my family back for everything they've spent on getting me out of here.
One day get married and start a family of my own.

On the other side of the line:

Request a transfer to Rome, where living conditions for prisoners with long sentences are better.
Prepare for my Supreme Court hearing and try to have the trial restarted from the beginning in a different venue.
Try to graduate from UW at a distance (even possible?).
Write.
Stay in touch with family and friends as much as possible.
Earn five years off my sentence for good behavior.
Get prison job as a cleaner, librarian, or grocery distributor.
Send earnings home to help pay my parents back.

Hardest was my life-imprisonment list. It was the same as the twenty-six-year list, except:

Stop writing letters home.

Ask family and friends to forget me?
Suicide?

The appeal had gone so well. Losing would be all the more devastating. I was afraid I might stop breathing in a claustrophobic panic. I wondered if I'd ever be happy again.

The appeal was my last chance. If I were to be condemned again, I didn't think the Supreme Court would exonerate me.

I knew that my mother's perpetual optimism masked her real feelings. She'd be even more distraught than I if I were convicted. I imagined her going home without me, completely broken, and I knew that, in that moment, when I couldn't be there, I would want somehow to comfort her.

> *Dearest Mom,*
>
> *I love you. I'm writing this letter in case you come home and I'm not there with you to receive it, just in case we didn't win and I won't be coming home for a long time.*
>
> *I want you to know that I'm okay. I love you and I know you love me. I'm okay because I'm not dead inside, I promise, and I don't want you to be dead inside. The shit we can't control, the things that make us suffer, challenge us to be stronger, give us the opportunity to survive and be stronger, smarter, better. We are the only ones who know just how much we and our lives are worth, and we must choose to make the most of every passing moment, no matter where we are.*
>
> *I've thought of ways to make my life worth it, and I want you to remember exactly what makes your life worth it. Don't be lost— don't lose yourself. Read, walk, write, dance, breathe, because so will I.*

Amanda Knox

I'll be seeing you tomorrow in court. I'm ready. I'll be paying attention and reflecting on what to say in the end. You'll have to tell me, now that it's over (by the time you receive this) what you thought.

I can't wait to see you. I love you so much.

Please hug Oma for me.

Remember it's only you who can make your life make sense. Thank you for always reminding me the truth about love.

I love you always,
Amanda

Chapter 35

October 3, 2011

I t was Verdict Day.

The numbers of press in the pit at the back of the courtroom and in the pressroom next door had steadily swelled.
My family had heard there were more than five hundred journalists covering the closing arguments and verdict, and they told
me that satellite trucks were parked six across in the piazza in
front of the courthouse. Their presence guaranteed that the announcement of a verdict—the most deeply affecting moment of
my life—would be beamed around the world.

Mom and Chris, Dad and Cassandra, Deanna, Madison, and
my aunts Janet and Christina were in the courtroom. Having
everyone there was huge. It was a show of force that let me know
I wasn't alone, that they loved me no matter what. For the last
four years, their lives had been on hold, too. My mom and stepdad, my dad and stepmom, and my grandmother had mortgaged
their houses to pay for everything from my groceries to my legal bills; from their shared rented apartment outside Perugia to
airfare back and forth to Seattle. They'd sacrificed everything
to make sure one of them was there during the eight hours a

month I was allowed visitors. My father told me how, over the dozens of drives he'd made to the prison, he'd watched the seasons change and the years go by. He'd passed the same farmland as it was tilled, harvested, turned under. He'd seen buildings go up from foundation to finish. Deanna had been so traumatized she'd dropped out of college.

Even so, they'd only ever been able to cheer from the sidelines. Of the 34,248 hours I'd spent in prison since November 6, 2007, I'd been allowed to see them for 376 hours—1 percent of the time.

Still, I needed them absolutely, whether the outcome was good or bad. If my life were definitively taken away from me, they'd be the only good I had. If I were acquitted and released, they'd be the ones I'd return home to. The decision would affect us all.

Around my wrist I was wearing a star I'd crocheted. I wore it to every hearing—not as a good luck charm but as a personal emblem. I'd made the star and many more like it for my family, early in my imprisonment. The thread, once a pristine white, had dirtied over the years. The star was my humble attempt to create something new and beautiful from what little I had available to me.

Judge Hellmann and Assistant Judge Zanetti were there, along with the six jury members wearing their Italian-flag sashes. The Kerchers wouldn't be there until later, for the verdict.

Raffaele spoke first, taking off his white Livestrong-type plastic bracelet reading "*LIBERA AMANDA E RAFFAELE*"—"FREE AMANDA AND RAFFAELE." It was a supporter bracelet made by my family. He said he'd worn it since our conviction. He held it up, an offering to the court in the hope that he wouldn't need it anymore.

"I have never harmed anyone," he said. "Never in my life."

My turn came next. I was shaking so badly the judge asked if I wanted to sit down. I hadn't eaten or slept in days, and tears came as soon as I started to speak. I was wringing my hands in front of me, pleading for my life.

"It was said many times that I'm a different person from the way I look. And that people cannot figure out who I am. I'm the same person I was four years ago. I've always been the same.

"The only difference is what I suffered in four years. I lost a friend in the most brutal, inexplicable way. My trust in the police has been betrayed. I had to face absolutely unjust charges, accusations, and I'm paying with my life for something that I did not do.

"Four years ago I was four years younger, but fundamentally I was younger, because I had never suffered . . . I didn't know what tragedy was. It was something I would watch on television. That didn't have anything to do me . . .

"I am not what they say I am. The perversion, the violence, the spite for life, aren't a part of me. And I didn't do what they say I did. I didn't kill. I didn't rape. I didn't steal. I was not there. I wasn't present at this crime . . .

"I want to go home. I want to go back to my life. I don't want to be punished, deprived of my life and my future, for something I didn't do. Because I am innocent. Raffaele is innocent. We deserve freedom. We didn't do anything not to deserve it.

"I have great respect for this court, for the care shown during our trial. So I thank you."

I sat down and silently sobbed. I'd never felt so small and insignificant. I was at the mercy of a court that had shown me no mercy for four horrible years.

Before the judge adjourned the trial, he warned the court,

"This is not a soccer game, a terrible crime has occurred . . . now the lives of two young people hang in the balance . . . When the verdict is announced, I want no *tifoseria*—'stadium behavior,'" he said. Then the judges and jury withdrew into chambers, and I was led from the courtroom.

Before being brought to the garage and locked in the prison van, I was allowed to hug my family in the back hallway. Raffaele was there, too, with his family. I asked him if he was nervous. "No," he answered. But it was a very tentative-sounding no.

When I got back to Capanne, Don Saulo greeted me at the entrance to the women's ward. "I've put everything off today to be with you," he said, taking my hand. "My office is completely at your disposal."

"Please, let's go there now," I said. Once there, I strummed the guitar and sang along to Mass songs that we both liked. Then we pulled out the keyboard, and I practiced the song I'd just learned—"Maybe Not," by Cat Power.

Don Saulo took out a pocket tape recorder. "Just in case I don't get to hear you sing again for a long time," he said, smiling.

I sang the song again. Soothed by Don Saulo, my voice was steady.

The rest of the time, we sat across the desk from each other, talking. As he'd done so many times before, he held my hands— and as always, it gave me comfort.

Don Saulo's parents had sent him to seminary when he was eleven. He'd been on his own most of his life.

"Are you lonely?" I asked. It wasn't the first time I'd asked, but he always deflected the question.

This time, he answered. "Yes," he said. "But I have God. It's a fulfilling existence, but it's also lonely. If you serve a certain purpose to humanity, humanity doesn't always serve you back. In *seminario* they almost prepare you for that by being really formal, so you don't get too connected to people. You're not allowed to have special friends."

"That makes me sad for you, but somehow you turned out to be such a strong, caring person."

After a few minutes of silence, I said, "I'm so scared."

This was not news to him.

"But I'm ready for whatever happens. I've thought it through. I've made lists. I've written my mom. I'm not going to let this destroy me."

"I'm going to be praying for you," he said, squeezing my hands, his cheeks wet. "I'm praying that you go home, Amanda. I hope I'll never see you in prison again."

"I'm really going to miss you if I'm freed."

I'd allowed myself the tiniest shred of hope to say those words.

Don Saulo gave me a good-bye present: a small, silver flying dove on a thin chain. "The dove represents the Holy Spirit for my church, *Santo Spirito*, and it also represents freedom," he explained.

Around 4 P.M., it was time for Don Saulo to leave. He hugged me for a long time. "I love you like a grandfather," he said, holding me.

"I love you, too, Don Saulo."

As I headed upstairs to my cell, an *agente* told me that Rocco and Corrado were waiting to see me. We detoured to the foyer of the women's ward, where Comandante Fulvio, the head of the prison, was talking cheerfully with them. They were both smiling. "Where've you been?" Rocco cried teasingly.

"How are you feeling?" asked Corrado, steering me into a private office.

"Really nervous," I said.

"We can understand your nerves," said Corrado, "but everything has changed since before."

"After your verdict, we've arranged for a car to pick you up from the prison," Rocco said. "So you're not swarmed by journalists."

"We'll both be here to take you to Rome," Corrado said. "We've worked it out with your parents. We're just finalizing the details with Fulvio."

Their plans seemed wildly overconfident.

"There's nothing to be afraid of, Amanda," Corrado said, squeezing my hand tightly.

Back in my cell, Irina and I watched TV. Every channel showed a crowd gathering and hordes of journalists outside the courthouse waiting to be called in to hear the verdict. The reporters recapped the last four years. I liked watching them talk about how the appeal had turned to favor the defense due to the independent experts who had poked gaping holes in the key evidence. Some thought we would win; others disagreed. When the latter came on, I changed the channel.

I reminded myself that none of them really knew anything. After all, they were reporting that on the chance I was freed, my family had hired a private jet to fly me home. They wouldn't have mentioned that rumor if they knew my parents were in no position financially to do such a thing.

It was about 8:30 P.M. when the *agente* came to get me.

I shook the whole way to the courthouse. New hives seemed to appear at every turn of the van.

The holding cage at the courthouse felt like a brick oven. I could hear Raffaele fidgeting on the other side of the thick wall. A couple of times he asked if I was all right.

"Yes," I said, my voice weak. My chest hurt. I was shivering uncontrollably. Violently, as if I had a fever of 105. I tried to take deep breaths. It didn't work.

The guard holding my arm was a tall, sturdy blonde who wore baby-blue eye shadow to match her light blue uniform. I whimpered like a child as we headed upstairs. "*Shh, shh,*" she said. "It's okay."

Ashley and Delaney were standing by the courtroom door, just as they had at the end of the first trial. The blond *agente* and a young ponytailed brunette half-pulled me inside. I wasn't ready to face the decision.

The courtroom had the feel of a carnival put on mute—an electric, unnatural silence emanated from so many varied bodies. The journalists were allowed to film the reading of the verdict, so there were flashing still cameras and TV cameras. Some journalists had hoisted themselves up on tall stands erected for this moment.

I didn't see Meredith's family coming in. I saw my family standing alert against the fence that ordinarily separated the journalists from the rest of the courtroom. They were smiling at me, but I could see the strain on their faces. I tried to smile back, but my face had frozen. I was pushed between my lawyers, who put their arms around me and tried to calm me. "Are you cold?" Luciano asked. He rubbed my shoulders, pulling me against him.

I didn't want the judges and jury to come out.

I didn't want to hear their words. I was too afraid they would condemn me.

The court secretary announced, "*La corte!*" and the court filed into position. The jury members looked steadily ahead, expressionless. Judge Hellmann began reading aloud. He started with the letter F, the charge for slander. I was pronounced guilty, and my sentence was raised from one year to three—time I'd already served.

I tried to hold off the terrible resurgence of fear that I'd be convicted on all charges. I had to focus to remember that I could be convicted for slander and not for murder, and I strained against the feeling of drowning until I knew what was coming.

People started muttering behind me, turning my nerves up another notch.

"For the charges prescribed in letters A, B, C, D, and E," Judge Hellmann continued, *"La corte assolve gli imputati, per non aver commesso i fatti"*—"the defendants are acquitted by the court, for not having committed the acts."

I'd been struggling for four years to reach the surface of the water. That first gasp of air was deep, heavy, sudden, and painful. But it was breath. It was life. It was freedom.

The crowd cheered. Some booed.

In the seconds before the guards rushed me out of the courtroom, Carlo, Luciano, and Maria hugged me. Then, in one step, I was out, down the stairs, crying, propelled by the guards, who guided my steps.

Waiting downstairs, I was still crying. Raffaele held my hand, saying, "But Amanda! It's okay! We're going home!"

One of the guards winked. "You've done well, little girl," he said. "You won me my bet!"

Just before we were taken outside, I embraced Raffaele. "I'll come see you in Seattle," he said.

The guards helped me into the backseat of a regular police car. I buckled myself in and looked out the windows at the trees and cars lining the street. It was dark, but I wanted to see everything—just to see them—because I could. We raced through Perugia and into the countryside, sirens blaring—just like when I'd been taken to the prison the first time. This time I didn't have my head down between my knees. This time the guards were smiling at me.

At Capanne, the paperwork had already come through from the courthouse. "It's time to go," the guards said. "Run upstairs quickly and grab your things. *Now*. Run."

I ran through the hallway alone, up the stairs to the locked gate. As soon as I called out, "*Agente!*" for the guard to let me in, the hallway erupted in cheers. Women crowded in front of their gates to see me, thrusting their arms through the bars to touch me, to say good-bye. I touched their hands, a passing marathon runner.

Irina was at our cell door crying. We hugged. "Take care of yourself," I said.

I picked up my bag, walked out of my cell for the last time, and ran back down the hall to shouts of "Amanda, Amanda. *Libertà! Libertà! Libertà!*"

Guards took me to the center building and gave me back my passport and money. I didn't recognize my passport photo. It was a younger me, taken just before I'd left for Italy, before any of this had happened, a picture of the Amanda who'd left Seattle in a rush, excited to be in a new place, to discover herself as an adult in a new culture. I felt sad to see her. I wanted to say, "You have no idea what's going to happen to you. I want to protect you."

I was handed a baggie with the earrings that guards had confiscated when I got to Capanne. The holes in my ears, new when I went to prison, had long since closed up.

Rocco and Corrado had caught up to me.

It was all happening so fast.

They hugged me, teary-eyed, saying, "We're going to get you out." They handed me a new iPhone, still in the box. I'd never seen one. "Use this to call us," Rocco said.

My stomach was turning over and over, my face hurt from smiling. My heart hurt from beating so hard. I was out! I was free! I was so overwhelmed, I couldn't talk. I hurt with joy.

I came out the same door I'd gone through four years before. I remembered to brush my right foot against the ground, the prison ritual to pass on freedom to another prisoner. Comandante Fulvio smiled and shook my hand. Prisoners erupted in cheers, banging pots and pans and waving T-shirts and towels through the bars of their windows, screaming, "*AMANDA! LIBERTÀ! E vaiiiiiiiii! LIBERTÀ! AMANDA! VAI A CASA! LIBERTÀ!* Yaaaay! Freedom! Amanda! You're going home!"

I was free.

I was going home.

October 3–4, 2011

C orrado got into the backseat of a black Mercedes with tinted windows and leather seats while I hugged Rocco for the last time. He would be staying behind with Comandante Fulvio. "Thank you for all you've done," I told him, my heart racing. Everything was in a rush.

"Go on. Go on," he said. "Call me later." He winked.

I got into the backseat of the car and shook the hand of the uniformed driver, a polite and quiet young man who couldn't have been much older than I was. "What's happening now?" I asked Corrado.

The driver started the car and pulled it to the gateway while Corrado quickly explained the plan. "Your family is waiting outside the gate. They'll follow us to Rome. We'll pick up your mom along the way. Once in Rome, we'll drop you off at a safe house for the night. You'll be flying home tomorrow!"

I wasn't prepared for the scene outside the car window as we crawled out of the gate. The driver pushed us through, but we were met with a wall of cameras and people. It was so dark out that I could see only from the flashes of the cameras—journalists surrounding the car, faces practically pressing against the glass. I automatically ducked my head, an instinct I had developed.

Once past the horde of journalists, the driver hit the gas and we took off down the road and into the countryside, in the opposite direction of Perugia's city center. I had never turned left going out on that road before, and only had a vague understanding that we were heading west-ish—I wouldn't be able to see it in the darkness, but I knew we were passing a hill on top of which was a farmhouse I had often seen during the past four years.

We were driving as quickly as possible, zooming down the curving streets as if in a high-speed chase. I looked behind me and saw a line of cars zooming after us. It *was* a high-speed chase. There were journalists, paparazzi. "You said we were going to pick up Mom along the way?" I asked Corrado incredulously. How could we go anywhere without being swarmed?

"Your mother should be in the car following immediately behind us," Corrado explained. "Our driver will try to lose these journalists and find a quiet spot to pull up and for her to get in."

Mom! I thought. I looked back again and saw the headlights of the car close behind us, leading the string.

"They're right on us," the driver said, referring to the journalists. He turned off the headlights so they would have a harder time seeing us, and we started taking fast turns off the road, circling. I couldn't see anything from the window. We had to do this maneuver a few times before we found a quiet, deserted patch of gravel, and the car that had been on our tail, headlights turned off as well, growled to an idling halt beside us.

I opened the door, and Mom hurtled into the seat next to me, so I ended up in the middle between her and Corrado. She was frazzled, but it took her only a split second before she burst into tears and embraced me. In the meantime, another figure hustled into the front passenger seat, the doors closed, and we were off again, kicking up gravel. The car Mom had gotten out of similarly sped off behind us.

"My baby! My baby!" Mom gushed, and groped my head and shoulders. I buried my face in her neck, almost unbelieving that she was there, that she was in a car with me, that we were racing out of Perugia, racing toward Rome. The driver turned the headlights back on, and we were soon spotted again by journalists chasing us. The car Mom had gotten out of, which she said was being driven by Chris, was meanwhile zigzagging across the length of the already narrow road behind us, trying to keep journalists from getting around him and near our car. At a certain point, Chris told me later, he was even rammed from behind.

The figure in the front seat of our car finished strategizing with the driver and then turned around in his seat. "Hi. I'm Steve Moore." He shook my hand and smiled. "We can get to the pleasantries later." He turned back around and kept an eye on the road as we roared down the streets, eventually merging onto the highway. He must have been around my mom's age, and had the build of a retired baseball player—strong, padded. He was actually a retired FBI agent who had presumed me guilty until he looked into the case at the urging of his wife. As a result, he became an advocate for my defense in the United States, and wrote online about the evidence from a professional investigator's perspective, criticizing the prosecution's claims and explaining why they were wrong. He had written to me in prison, and I had written him back to provide ideas for the name of his daughter's new pet pig—my favorite idea, even if it was obvious, was inevitably Hamlet.

When Mom and I finished embracing, we clung to each other's hands as if for dear life. It had been years since I had seen her anywhere but in the prison visitation room and the courthouse. It almost seemed like the setting couldn't be real—more like a revolving two-dimensional backdrop on a stage.

I wasn't used to being in different environments anymore. Indeed, in retrospect, I think my memories of the four years I spent in prison are so clear precisely because the background never changed. It was always the same echoing hallway; the same bleached cells; the same desolate yard; the same dark, windowless innards of the prison van; the same bright, crowded courtroom. These settings served as the blank piece of paper upon which the changes in character and feeling were more starkly revealed.

These completely new things—a changing, moving landscape; a car with windows; my mother; my freedom—were overwhelmingly overstimulating. I was bouncing up and down in my seat with pent-up excitement. I hadn't felt so much energy in years—not within the prison, in which I'd so often felt lethargic because of my sadness and the emptiness of purpose I tried so hard to fill on my own.

I wanted to know how the family was, how everyone had reacted, where everyone was, if we would be able to meet up with them soon.

Mom, still crying, told me that as soon as the verdict was pronounced, everyone burst into tears and started hugging. "Of course," she said, "getting out of there was a mess. The journalists were going crazy, pushing each other over to try to get interviews. Your sister made a beautiful statement outside the courthouse. You'd be so proud of her."

I was proud of her.

"Do you want to call them?" Mom asked, excitedly offering her BlackBerry to me.

I didn't hesitate a moment, but I fumbled with the device, unable to make it work. I couldn't figure out how to get to the contacts.

"It's touch-screen, honey." Mom laughed. She scrolled through the phone. I hadn't picked up a cell phone in years, and never a touch-screen. This device was as good as sci-fi to me. But more than that, I was struck by how automatic it was to place a simple call. In prison, I had had to make a request a week ahead of time, ask to be allowed into the booth when the time came, wait for the prison official listening in to place the call, and say as much as I could in the ten minutes before the same official dropped the line. I recognized the names on the contact list my mom was scrolling through. I could now call any one of them, and talk for however long I wanted!

"Who do you want to call first?" she asked me.

I ended up spending the rest of the ride to Rome making call after call, to family and friends both back home and in Italy. I squealed loudly into the mouthpiece, too excited to keep my voice calm and the volume appropriate for the interior of a car.

By the time we pulled off the highway and into Rome, the paparazzi had long since lost our trail. I looked out the windows to view the city through the darkness, but my eyes kept falling on Mom, who was still clutching my hand and touching my cheeks. We drove through the streets, finally pulling up in front of an old town house on a street of town houses.

Steve got out of the car and surveyed the empty street before motioning us to follow. In the meantime, Corrado and the driver unloaded our bags. "We'll be back tomorrow to take you to the airport," Corrado said, smiling.

"Thank you so much for doing this," I said, hugging him.

"Are you kidding? I haven't felt so much excitement in a long time."

Steve led Mom and me up into the town house, where our host, a quiet, older man—a supporter of both mine and Raffaele's—greeted us with gestures more than words. He pinched my cheeks and showed us upstairs into his tight quarters—Mom and I would sleep on a foldout bed in the study; Steve on a cot in the kitchen. The rest of my family was staying in a hotel.

Mom and I put our bags down and headed for the bathroom together. I was carrying the toothbrush I had used in prison, but then remembered the insistence of my fellow prisoners that I was to break it and throw it away, to sever my last bonds with the prison and ensure that I wouldn't return. "Do you have an extra toothbrush?" I asked Mom.

"Of course. Let me go get it."

When she had gone, I wrenched at the toothbrush, managing to bend the handle. Good enough. I flung it in the trash.

Mom returned with a new toothbrush for me, and toothpaste, which we both shared. It was strange, to stand there brushing our teeth together, and made even more awkward precisely because we couldn't say anything with our mouths full. We met each other's eyes instead, arching our eyebrows up in recognition. This was like home again.

Mom was exhausted and crawled into bed immediately. But that first night, I couldn't sleep. My heart was still pounding. I wasn't remotely tired. I got up and slowly walked around the study, trying to read the titles of the books cramping the shelves around the walls. It was surreal to be in a place like this, when only hours before I had been sitting on my bed in prison, quivering with nerves and uncertainty.

I listened to Mom sighing in her deep sleep the entire night as I sat in an armchair, staring out the window until it became light. *This isn't prison*, I marveled to myself. *This isn't prison!*

Just after sunrise Steve was up and dressed, gathering our baggage in the foyer. He asked how I'd slept.

"Not at all!" I replied cheerfully.

Our host returned from his morning errand, offering us pastries and espresso. I gulped down the coffee and some water, but couldn't eat. I wasn't hungry. He then offered me the morning paper, tears in his soft eyes.

The front page showed a large, zoomed-in picture of me being escorted from the courtroom after the pronouncement of the verdict, my face contorted by crying. *Oh God*, I thought.

"You're so photogenic," Steve wisecracked.

We barely had time to gulp down the espresso before Corrado and the same driver from the evening before pulled up outside. After a long, reflective night, the rush was starting again. We said a quick good-bye to our host, who embraced us all with more strength than I would have expected from him.

We had some time before we needed to be at the airport, so we decided to stop by my family's hotel. Steve thought it would be a good idea if I changed clothes. "No offense. You look nice," he said, "but the paparazzi are looking for you in that outfit. We'll have a better chance of avoiding drawing attention to ourselves if you can borrow something from Madison or your sister."

It was early enough that when we pulled up outside the hotel there wasn't yet any movement around the building or in the lobby. Steve got out of the car first and looked around furtively. "We shouldn't stay here long," he said. "The paparazzi will have followed us here."

I pulled the hood of my coat up over my face and quickly approached the entrance to the lobby. I put down my hood and walked quickly toward the elevators with Mom and Steve, trying to look casual. Just then, a photo of me came up on the large-screen TV in the lobby, announcing my release from prison. I was still wearing the same clothes. We raced to the family's rooms, where we quickly greeted, squealing; and I changed clothes. Then we were off again.

Corrado had arranged for us to wait in the airport's VIP Lounge, and for even more than that. We drove in through a side entrance to the airport, went through private security, and, when we finally had to come out to face the public, had an escort through the airport hallways. We were a large group, walking quickly through the various terminals. But airport security surrounded Corrado, Mom, Steve, and me all the way to the lounge. In the meantime, people took out their iPhones and took pictures. I embraced Corrado.

My family gathered in the lounge. I ordered my first legitimate cappuccino in years, as I explained to my baffled little sisters how, in prison, to create foam for a cappuccino, we'd pour boiling hot milk into an empty two-liter bottle and shake it furiously. It never worked all that well.

Chris and a supporter who worked for the airline had secured us three business-class seats, so that I would be safe to relax on the plane, and family could be up there with me. Many journalists had managed to book last-minute seats on the flight, but the flight attendants, alerted to our situation, kept them from approaching the upstairs part of the plane.

Once on the plane, Deanna and I giggled together like little girls, poking at each other across the armrest. She fell asleep, but

I still couldn't even doze. I was wired. I spent a lot of time catching up in my mind with what had happened to me so far, since my acquittal. Everything was different. Everything was not prison. I had viewed the same countryside for years without ever seeing anything else, and now I could open my window shade and watch the clouds foam beneath us. The flight attendants were smiling and considerate. I watched Deanna sleep, curled up awkwardly even in roomy business class. Chris was across the aisle.

I searched for something to watch on the video screen on the seatback in front of me, and came upon a news report about my case. I felt myself go numb. This was a British report, and I was suddenly reminded of how big this was, of how many people knew about this and were following what was happening. I watched myself being led from the courtroom after my acquittal, and immediately my chest tightened. I took off Chris's headphones and turned off the channel, struggling to breathe.

About an hour before we were due to arrive in Seattle, Mom came up from coach to switch seats with Deanna. She brought with her little notes from journalists on the plane. "They all seem very supportive," Mom said, after we read the notes together. All were congratulatory in nature, and all were requests for interviews.

The sight of coming down over Seattle jacked my heart rate up again. I plastered myself against the window, almost in disbelief at how familiar the view was. It was so familiar that, for a brief moment, it felt as if I hadn't been gone so long after all, as if my life hadn't been irrevocably changed.

But after the plane had landed, it was back to the crazy rush and the new reality of the present. Coming out of the doorway of the plane, I was hit immediately by the smell—the wet earthiness of Seattle, especially in the fall, which was so completely different from the way Perugia smelled at any time of the year. I breathed in deeply. It was the first thing that truly made me realize I was home.

Our group was allowed to exit the plane first, and down onto the tarmac, where two large black vans were waiting for us. The vans drove us to a private area, where we were greeted by Seattle police officers. My heart clenched uncomfortably for a moment, but they smiled. They were there at our service, to escort us.

There, I met David Marriott and Ted Simon for the first time. David helped my family handle the media, and Ted gave my family legal counsel. I recognized them both immediately, from the descriptions of them my family had pieced together for me. David looked like Santa Claus. He smiled and wrapped his arms around me, pulling me up against his belly, his voice joyful. "It's so good to finally meet you!" he cheered. Ted looked like a smartly attired, wily fox. He had thick, wavy gray locks and was wearing impressive black cowboy boots with his well-fitting suit. We also embraced, and he spoke energetically. "It's such an honor! How're y'doing? How y'feel? It's so great!"

David and Ted explained that there was a press conference about to happen outside. Ted and my parents were going to say a few words and take questions. "Now, if you think you can do it, you may also give the press a little something to sit on," David said. "Then maybe they'll be more inclined to leave you in peace for a while. Something like that?"

I felt it was the right thing to do. We walked through the doors and onto a dais that had been set up in front of a bank of cameras. My imprisonment and release from prison had never been a private affair. I gathered myself, feeling that I would be able to pull through without getting choked up.

I had expected to confront the same sort of impersonal cameras I had shied away from in the courtroom. Instead I was received with cheers. I was blown away by the show of support.

I said what first came to mind, often pausing to try to think of what to say next.

> *"My family is reminding me to speak in English, because I'm having problems with that. I'm really overwhelmed right now. I was looking down from the airplane and it seemed like everything wasn't real. What's important for me to say is just thank you to everyone who has believed in me, who has defended me, who's supported my family . . . My family is the most important thing to me right now, and I just want to go and be with them, so, thank you for being there for me."*

I couldn't stay focused, and ended up backing off from the dais. I could have said so much more about the many lives forever marked by the events of the past four years. I wish I had said something about Meredith and her family. I wish that I had expressed things more eloquently. But I had been unprepared. I was unprepared for my freedom. I was unprepared for how unfamiliar this all felt.

And still, I was home.

Author's Note

The writing of this memoir came to a close after I had been out of prison for over a year. I had to relive everything, in soul-wrenching detail. I read court documents and the transcripts of hearings, translated them, and quoted them throughout. Aided by my own diaries and letters, all the conversations were rendered according to my memory. The names of certain people, including friends, prisoners, and guards, have been changed to respect their privacy.

So much has been said of the case and of me, in so many languages, in so many books, articles, talk shows, news reports, documentaries, and even a TV movie. Most of the information came from people who don't know me, or who have no knowledge of the facts.

Until now I have personally never contributed to any public discussion of the case or of what happened to me. While I was incarcerated, my attention was focused on the trial and the day-to-day challenges of life in prison. Now that I am free, I've finally found myself in a position to respond to everyone's questions. This memoir is about setting the record straight.

I've written about what brought me to Italy, how Meredith's murder affected me, and how I got through imprisonment and a judicial process married to the media. I went in a naïve, quirky twenty-year-old and came out a matured, introspective woman.

I'm grateful to Robert Barnett, for graciously and securely holding my hand along this new, unfamiliar journey through publishing and the public.

I'm grateful to my publishing house, HarperCollins, for giving me the opportunity to be heard. Tina Andreadis, Jonathan Burnham, Michael Morrison, Claire Wachtel—thank you for believing in me, for supporting me, for your ideas and your care.

I wouldn't have been able to write this memoir without Linda Kulman. Somehow, with her Post-it Notes and questions, with her generosity, dedication, and empathy, she turned my rambling into writing, and taught me so much in the meantime. I am grateful to her family—Ralph, Sam, Julia—for sharing her with me for so long.

I can only attempt to recognize and thank all the sources of support my family and I received over the course of our battle for justice in Italy. I'm grateful to everyone who gave their time, their words, and their means to support us.

Don Saulo and the prisoners of Capanne, who appreciated me for who I am, supported me through many a moment of crisis, and taught me so much about humanity.

Rocco Girlanda and Corrado Daclon, for visiting, for supplying the books and music to keep my mind active during imprisonment, and for the help they provided getting me home from Italy.

Dr. Greg Hampikian, Dr. Saul Kassin, and Steve Moore for their advocacy, their expertise, and their friendship.

David Marriott and Theodore Simon, for their guidance and generosity.

Professor Giuseppe Leporace and the Seattle Prep community, for their dedication as educators to my mind and heart, despite the criticism.

My family and friends, for coming together in my time of need, for overcoming the unknown, for saving my sanity and my life.

And finally, Luciano Ghirga, Carlo Dalla Vedova, and Maria Del Grosso, for defending and caring about me as if I were one of their own.

About the Author

Amanda Knox was convicted of murder in Perugia, Italy, in 2009. In 2011, the conviction was overturned, and she was acquitted of the charge of murder. She now lives in Seattle, her hometown, and is studying creative writing.